The Changing Face of Land and Conservation in Post-colonial Africa

The year 2013 marked the 100th anniversary of the 1913 Land Act in South Africa which legalised the violent dispossession and alienation of the African majority from the land. It is common cause that the alienation of land for conservation purposes, introduced to Africa under colonial rule, has continued more or less uninterrupted until today. However, while nature conservation practices inevitably raise challenging questions relating to land and land use, there has thus far been little concentrated effort to bring together scholars working on the land question, particularly around issues of land tenure, with those whose work focuses on questions of nature construction and the social impacts of conservation in an African context.

Compiled from research presented at a ground-breaking interdisciplinary conference held at Rhodes University in Grahamstown, South Africa, in 2012, the chapters in this book made their first appearance in a special issue of the *Journal of Contemporary African Studies* (JCAS) in July 2013. The book brings critical interdisciplinary analyses of the complex inter-relations between contemporary (neoliberal) conservation practices in post-colonial Africa, into conversation with the well-trodden territory of land use and contested land issues on the continent. Anchored by an intellectual curiosity about the extent to which past practices continue into the present and with what consequences, the book provides fresh insights into the complex relationship between land and conservation in contemporary Africa.

George Barrett works in the field of international relations, security and environmental politics. Her doctorate argues for the reinvigoration of Critical Security Studies in Southern Africa, giving centrality to human-environment relations in rethinking security and emancipation in the region. Her research views these issues through the optic of peace parks.

Shirley Brooks is a human geographer and environmental historian. Her research explores social histories of nature conservation and land issues in South Africa. She has written on the history of the Zululand game reserves as well as private wildlife production and related geographies of tourism and hunting in KwaZulu-Natal.

Jenny Josefsson's research explores the relationships between people, land and wildlife on private game farms in KwaZulu-Natal. She has a special interest in the dynamics of ethnographic fieldwork and knowledge-generation.

Nqobile Zulu's research interests span across the spectrum of Agrarian reform, development and the interface it brings to rural communities. His current work focuses on the transition to private game farming and the contestations developing in the Groot Marico district of the North West province in South Africa.

The Changing Face of Land and Conservation in Post-colonial Africa

Old Land, New Practices?

Edited by
**George Barrett, Shirley Brooks,
Jenny Josefsson and Nqobile Zulu**

LONDON AND NEW YORK

First published 2015 by Routledge

2 Park Square, Milton Park, Abingdon, Oxon, OX14 4RN
605 Third Avenue, New York, NY 10017

Routledge is an imprint of the Taylor & Francis Group, an informa business

First issued in paperback 2020

British Library Cataloguing in Publication Data
A catalogue record for this book is available from the British Library

ISBN 13: 978-1-138-83273-2 (hbk)
ISBN 13: 978-0-367-73894-5 (pbk)

Typeset in Times New Roman
by RefineCatch Limited, Bungay, Suffolk

Publisher's Note
The publisher accepts responsibility for any inconsistencies that may have
arisen during the conversion of this book from journal articles to book chapters,
namely the possible inclusion of journal terminology.

Disclaimer
Every effort has been made to contact copyright holders for their permission to
reprint material in this book. The publishers would be grateful to hear from any
copyright holder who is not here acknowledged and will undertake to rectify
any errors or omissions in future editions of this book.

Contents

CONTENTS

Citation Information

The chapters in this book were originally published in the *Journal of Contemporary African Studies*, volume 31, issue 3 (July 2013). When citing this material, please use the original page numbering for each article, as follows:

Please direct any queries you may have about the citations to
clsuk.permissions@cengage.com

Introduction

It is particularly apposite that this special issue on 'Old Land, New Practices' should come out this year. Exactly a century ago, the notorious 1913 Natives' Land Act was passed confirming in law the plunder of the colonial wars of dispossession. It reserved a mere 7.3% of the land surface in South Africa for the exclusive occupation of the African majority, thus claiming the overwhelming bulk of the country for white ownership and control. The racialised splitting of the country has had long-lasting spatial consequences as this old land law is inscribed in the current context within which new practices are struggling to emerge. This special issue thus allows us to reflect afresh on the enduring reality of this vital piece of segregationist legislation.

The articles in this special issue are from a Conference held at Rhodes University in Grahamstown in September 2012. Both the town and the university remain marked by colonial and apartheid geographies. For example, the main entrance to the university, the Drostdy Arch, in the not too distant past represented a gate of racialised exclusion as it was once the entrance to a military garrison. The Conference commenced on 12 September, the anniversary of Steve Biko's death and it is important to recall that Rhodes University had a lot to do with his decision to launch South African Students' Organisation (SASO) and the black consciousness movement more generally. In many ways his black consciousness journey took a crucial turn at Rhodes at a NUSAS (National Union of South African Students) conference held on campus in 1967, when the University prohibited the black delegates from sharing residence accommodation with white delegates in line with apartheid prescriptions. Biko proposed a motion to adjourn the meeting at Rhodes and to reconvene at a township venue where the white and black delegates could meet and be accommodated together. The motion was defeated and consequently, Biko and his colleagues walked out having had their suspicions of the duplicitous role of white liberal students confirmed. Inasmuch as Rhodes had something to do with his black consciousness trajectory, his journey of life ended in Grahamstown 10 years later when he was arrested after returning from a failed attempt to meet with the late Neville Alexander in Cape Town in an effort to unify the various strands of the liberation movement. Soon after his arrest 35 years ago to the day of the commencement of the Conference, he was dead and South Africa lost one its most outstanding thinkers and leaders.

Land and race lie at the heart of the enduring problems of imagining a unitary conception of the South African nation. Thus, in the same year that Steve Biko was killed, Basil Kivedo[1] could say, 'Besoedeling is 'n wit man se probleem' (literal translation – pollution is a white man's problem).

His statement that pollution and by implication matters related to the protection of the environment are the exclusive domain of whites was informed by the assumption that the major preoccupation for blacks should be the battle against apartheid and for social and economic equality. Conservation and nature were luxuries that blacks could ill afford to take up as issues. Connections between concerns for conservation on the one hand and the struggle for political rights on the other were as wide apart as the racialised divisions of apartheid. In fact they mirrored those divisions. In reflecting on this gulf of activity and priority, this Conference is a very important intervention demonstrating the intimate linkages between the environment and political rights and suggesting a much more nuanced appreciation of the dialectical relationship between human and non-human nature.

The conference sought to highlight some of the continuities between colonial and post-colonial land use management practices. Some years ago, Gavin Williams promised us a manuscript on the similarities between the villagisation scheme of UJAMAA in independent Tanzania and the very unpopular Betterment Programme and Rehabilitation Schemes in South Africa of the late 1930s and early 1940s. Unfortunately, this book project was abandoned in favour of his work on Wine farms in the Western Cape – but the point was well taken. Williams was concerned with the manner in which independent African states followed the examples of their colonial predecessors. The presumption of much nationalist historiography in Africa is that struggle for national liberation was the antithesis of colonialism. In contrast, radical critiques have shown that in reality once independence was accomplished it turned out to mimic colonialism in very many ways. In respect of land, very few African countries embarked on a thoroughgoing process of land reform. One of the few which paid serious attention to land and agrarian relations was Ethiopia, after the collapse of the Haile Selassie regime. However, today the Ethiopian government is at the forefront of selling off large portions of its land surface to all and sundry, but specifically to multinational corporations.

Land grabbing is unquestionably one of the greatest challenges facing sub-Saharan Africa in the last decade or so. More than two-thirds of the land grabbed globally has been in Africa and there are a wide variety of reasons for this related to the nature of land rights of the people and the nature of the state. But it is also connected to perceptions that development could be brought to the continent through huge corporate means. Some of the very same firms which were responsible for the global economic crisis are also responsible for the alienation of large tracts of land across Africa. Land grabbing takes many forms, but one of its critical features is the complicity of the host government in various deals, from outright sales to long-term leases. Precisely because of the involvement of governments in these deals, it is almost as if there is no need for regulation – the firms have been given carte blanche to virtually do as they please, as long as the measly rent is paid to the government. Of course, there is invariably no compensation to the mass of the people for the loss of their land. Its results also vary from the displacement of small-scale farmers to environmental destruction and loss of water. Water is a critical issue, as much of the land grabs revolve around the right to appropriate water resources.

Spatially in South Africa the old and the new continue to compete for legitimacy. A new geographic dispensation divides the country into nine provinces. Yet in reality, the spatial divisions of the past have proven to be much more durable than was initially assumed. The homelands or bantustans have been formally disbanded, but these areas remain differentiated from the rest of South Africa by a distinctive form of land tenure and a different form of local government, one that recognises the role of chiefs. And so, when Mamdani argues that the main problem for all national liberation movements across Africa was that they did not have an agenda for democratising customary power, he was talking about SA as well. This year, in its submission to the constitutional review committee, the National House of Traditional Leaders proposed a constitutional amendment which would remove sexual orientation as grounds for non-discrimination.

The articles in the special issue are very well introduced by George Barrett, Shirley Brooks, Jenny Josefsson and Ngobile Zulu. They deal with three main themes. Firstly, they question the manner in which the notion of community is constructed. In respect of South Africa, the spatial divisions in land between communal areas and freehold areas denote a very real distinction reminiscent of Mamdani's bifurcation of citizens and subjects. The very idea of a traditional community has crept into official discourse where it assumes increasing legitimacy. Yet it is profoundly problematic from the perspective of democratic citizenship. Secondly, they try to highlight the many hidden voices of ordinary people as they struggle for survival in contestation over conservation space. Finally, they provide an overarching critique of the dominant neoliberal approach which has the effect of commodifying conservation.

Note

1. Then Sociology lecturer at the University of Western Cape and currently the Mayor of Worcester in the Breede River Valley Municipality.

Fred Hendricks
Rhodes University, Grahamstown, South Africa

Starting the conversation: land issues and critical conservation studies in post-colonial Africa

George Barrett[a], Shirley Brooks[b], Jenny Josefsson[c] and Nqobile Zulu[d]

[a]Department of Political and International Studies, Rhodes University, Grahamstown, South Africa; [b]Department of Geography and Environmental Studies, University of the Western Cape, Cape Town, South Africa; [c]Geography Department, University of the Free State, Bloemfontein, South Africa; [d]Department of Development Studies, University of Witwatersrand, Johannesburg, South Africa

This thematic issue brings together the scholarly fields of critical conservation studies and African land issues, a relationship largely unexplored to date. The alienation of land for conservation purposes, introduced to Africa under colonial rule and still taking place today, has fundamental impacts on the politics of land and land use, and is contested in contemporary nation-states – including those that are attempting to implement land restitution and reform. The contributors explore these issues in a range of African contexts. Three key themes are identified: the problematic constructions of 'community' by outside agencies; spatial exclusion and the silencing of local voices; and the neoliberalisation of conservation spaces. In contributing to new perspectives on these themes, this thematic issue shows how discourses and practices of conservation, increasingly shaped by neoliberalism, currently impact on land ownership, access and use. It further highlights some important historical continuities. These trends can be observed in transfrontier conservation areas, on state-owned land used for conservation and 'green' initiatives, but also on private land where conservation is increasingly turned to commercial purposes.

Land and conservation in post-colonial Africa: Old Land, New Practices?

This thematic issue of the *Journal of Contemporary African Studies* highlights recent scholarship on the complex interrelations between contemporary conservation practices in post-colonial Africa, in conversation with the well-trodden territory of land use and contested land issues in the continent. Within the last two decades or so, a substantial cross-disciplinary literature has emerged that engages with the local and regional politics of nature conservation from a social science perspective (see Anderson and Grove 1987; Neumann 1998; Zerner 2000, 2012, also see Marnham 1980 for an early journalistic critique). Critical conservation studies is emerging as a field in its own right, as evidenced by the success of recent conference initiatives attracting a range of scholars from various disciplines (see for example Nature™ Inc.). African land issues have been a subject of interest for Africanists over a far longer period, as scholars have grappled with the implications of communal tenure systems, land dispossession and the introduction of private property regimes in southern Africa's settler colonies (Allan 1965; Palmer and Parsons 1977; Bassett and Crummey 1993; Berry 1993; Evers, Speirenburg, and Wels 2005; Derman, Odgaard, and Sjastaad 2007; Benjaminsen and Lund 2012).

While nature conservation practices inevitably raise challenging questions relating to land and land use, there has thus far been little concentrated effort to bring together scholars working on the land question with those whose work has focused mainly on questions of nature construction and the social impacts of conservation in an African context. A conference held in Grahamstown in September 2012 was conceptualised with the aim of bringing together these two groups of scholars – as well as conservation practitioners – in productive, if at times challenging debate. Titled: 'Old Land, New Practices? The Changing Face of Conservation and Land-Use in Post-Colonial Africa', the meeting was jointly organised by a group of academics and postgraduate students with support from Rhodes University, the University of the Free State and the University of the Witwatersrand. The articles in this thematic issue were first presented as papers at this conference. The scope of interest includes the maintenance and extension of existing state-run conservation areas, new land allocations for conservation, the impacts of various forms of 'community-based conservation' and associated land-use controls in communal areas; as well as conservation enterprises on privately owned land and the impacts and outcomes of land reform.

Before introducing the articles in more detail, it is useful to reflect briefly on the focus and significance of the theme under review. The title of the conference implied a question with which contributors grappled – that is, the extent to which 'new' practices reflect discontinuity and a departure from the past, and whether qualitative shifts in practice are actually taking place or not. It is common cause that the alienation of land for conservation purposes, introduced to Africa under colonial rule, has continued more or less uninterrupted until today – albeit with the participation of different actors, under different circumstances and employing different pretexts. The many continuities in nature conservation, and the difficulties of 'decolonising conservation' in contexts where essentially colonial views of nature and its role still prevail in much policy and practice, are now quite well documented (see Adams and Mulligan 2003; Brockington and Igoe 2006). In southern Africa, the persistence after the end of apartheid of structures that govern conservation, as well as the people who are represented in these structures, is one example (see Draper, Speirenburg, and Wels 2004).

Other work, however, does suggest a qualitative shift. This thematic issue highlights a growing body of research showing how the discourses around nature, protected areas and wildlife have changed alongside a general mainstreaming of 'green' issues (Brockington and Duffy 2010; MacDonald 2010). In market-oriented approaches to conservation, a protected area is made to 'pay for itself' by identifying a niche and a value of some sort within a global marketplace increasingly well stocked with marketable nature and environmental initiatives (Büscher and Arsel 2012; Igoe, Neves, and Brockington 2010). This approach finds support in discourses of sustainable development and ecological modernisation, widely incorporated – at least at the level of rhetoric – into political and corporate agendas (MacDonald 2010). Such trends have developed as a result of (and in conjunction with) the neoliberalisation of political and economic ideologies and practices in the past four decades, where nature and conservation too are undergoing a process of neoliberalisation. This represents a significant shift in the ideologies of nature conservation and environmentalism, where 'nature' was originally conceptualised as a sphere requiring protection from economic forces (Brockington and Duffy 2010; also see Ramutsindela and Shabangu 2013, this issue). Utilisation and privatisation of natural resources, as well as the idea of a conservation area as a 'resource', are now normalised into policy and management practices. The commodification of nature is viewed as necessary for sustaining its existence, and *vice versa* – the use of nature and natural resources is proposed as a key to ending the current financial crisis (Igoe, Neves, and Brockington 2010).

The increasing commodification of life in all its forms is also reflected in the proliferation of 'transfrontier parks', known in the southern African context as 'peace parks'. These extend the reach of conservation practices into land that may formerly have been under different governance structures, and in the process create new security apparatus for protecting land. The development of southern African peace parks has provoked a growing inter-disciplinary critique. One focus is the implications of their development for state sovereignty and security in a region with a troubled history of inter-state relations (Duffy 2001, 2007; Ramutsindela 2007). It is not coincidental that the parks are located in border zone spaces where states have always had to assert their authority and delineate control (see Barrett 2013, this issue). Peace parks are ascribed the potential to bring greater security to the region and its people, particularly people who have historically and geographically been marginalised and located on the periphery of the state. However, the preoccupation with state-centric security concerns means that the security implications for actors other than the state are frequently obscured. Moreover, the pursuit of such aspirations within a neoliberal framework seems to (re)produce and exacerbate past experiences of insecurity and exclusion.

One way to bring the field of land and agrarian studies into closer proximity with critical conservation studies is for scholars in the latter field to pay more detailed attention to the way land tenure regimes influence and shape conservation practices in different contexts. Land held under indigenous or 'communal' tenure, or by land trusts, enables the promotion and implementation of various forms of conservation 'partnership' between state and private actors and so-called 'communities' (see for example, Ngubane and Brooks 2013, this issue; Godfrey 2013, this issue). Interventions by the state in land tenure (generally through various land reform initiatives, for example in post-apartheid South Africa) create new contexts for conservation practices such as co-management. While land held under freehold tenure is arguably less accessible than communal land to state conservation policies – although in the post-colonial context of southern Africa this ownership does not always go unchallenged – conservation authorities have embarked on stewardship programmes with the aim of attracting private landowners to adopt conservation-friendly management practices (for examples from South Africa, see CapeNature and Ezemvelo KZN Wildlife websites). These strategies aim to secure land for conservation in contexts where the state's options for acquiring new land are becoming more limited – at the same time, providing poor people with a viable land-use option, at least in theory.

More recently, attention has been drawn to the worldwide phenomenon of land grabbing, where often communally owned lands in the global South are acquired through a range of means (corrupt and otherwise) by large multinational corporations for commercial purposes. These include mining, the production of biofuels and food production for citizens of land-hungry northern countries (see Hall 2011; Borras et al. 2011; Borras and Franco 2012). Obviously colonialism itself was an ambitious and far-reaching form of land grab, and it is interesting to reflect on the differences between colonial 'land grabs' and today's version. Of particular interest to the theme of this special issue is the aspect of the so-called land grabbing phenomenon characterised as 'green grabbing'. Emerging scholarship on 'green grabbing' resonates strongly with the neo-liberalisation processes mentioned above in the context of new conservation practices. In a definition recently offered by Fairhead, Leach, and Scoones (2012, 237), the term 'green grabbing' describes contemporary forms of the 'appropriation of land and resources for environmental ends'. What makes it qualitatively different from past forms of colonial environmentalism, in their view, is that it 'involves novel forms of valuation, commodification and markets for pieces and aspects of nature, and an extraordinary new range of actors and alliances' (Fairhead, Leach, and Scoones 2012). The phenomenon of 'green grabbing' is thus gaining increasing prominence as a form of land and resource grab that involves 'new forms of appropriation of nature' (Fairhead, Leach, and Scoones 2012), leading to the restructuring of agrarian social and economic relations in Africa and elsewhere.

Introducing the articles: key themes and critical debates

The nine articles in this thematic issue, written from different disciplinary viewpoints, have important contributions to make to the debates highlighted above. As far as geographical range is concerned, the authors have worked in South Africa, Zambia, Uganda and Mauritius, suggesting common themes that extend well beyond the borders of a single country. Three broad groupings of papers are identified here, even though these themes are of course overlapping and some articles address more than one theme.

The first grouping of articles focuses on the often troubling politics through which 'communities' are constructed by conservation authorities, donor organisations and others involved in partnerships and (co)management of conservation areas as well as in land reform. A second set of articles speaks to the imposition of conservation management in particular spaces and issues of denied or excluded spatialities. Part of this story is the disallowing of resource use practices considered inappropriate under environmental governance regimes, and the effective silencing of dissenting local voices. 'Hidden' contestations over land and land use in these contexts are brought into the open, demonstrating how power operates in the exertion of control over land. The last group of papers deals most explicitly with the related theme of the neoliberal reconceptualisation of African nature and its (expanding) protected areas. These articles address questions around the neoliberalisation of conservation and environmentalism, and the commodification of nature through the positioning of transfrontier parks and other protected areas in the 'nature market' as well as so-called green economy initiatives such as carbon sequestration.

Constructing the 'community'

Godfrey's article paints a disturbing picture of top-down approaches to resource conservation and the ongoing construction of a 'mythic community' on the borders of South Luangwa National Park, Zambia. Her research, recalling many similarly top-down development programmes in Africa, reveals continuities between colonial and post-colonial discourses organised in fields of power that work to deny the rural poor not only their past but also, in a sense, their future. The article provides a powerful critique of the waves of apparently 'newly minted' community-based conservation initiatives imposed on local people by donor organisations, all taking a similar 'blueprint' approach and equally blind to the specificities of particular groups and histories.

This process of homogenisation and misrepresentation by outside promoters of conservation is also evident in Ngubane and Brooks' research on the creation of new 'community game farms' through the land reform process in KwaZulu-Natal, South Africa. In similar fashion to Godfrey, they address the complexities of 'community construction' and link this to a biodiversity conservation discourse that (re)constructs the identity of land beneficiaries as a single homogenous group. Their article further draws attention to the new power relations that arise from land restitution and the impact on land beneficiaries. Game farming is itself a 'new practice' on 'old land', as many freehold farms in different parts of southern Africa move from conventional areas of agricultural land use such as cattle farming, to wildlife production (in this case hunting farms) – a trend that has shown significant growth in the South African countryside in the last few decades (see Snijders 2012). The emergence of the 'community game farm' is a new twist to this tale.

Den Hertog's article on the !Xun and Khwe groups in South Africa is likewise sensitive to the construction of 'community', again in the context of land reform. Den Hertog examines the impacts of an involuntary amalgamation of two distinct groups, the !Xun and Khwe, into a uniform 'community' with one collective identity. One consequence of this construction – first created in the 1960s and since then reinforced a number of times – is an imposed sharing of space, most recently resulting from land redistribution. Like Brooks and Ngubane, den Hertog is concerned that current attempts by outside role-players to define 'community' identities, needs and aspirations continue to rely on colonial and apartheid classifications and stereotypes.

Spatialities of exclusion and silenced voices

Olivier's article draws attention to voices generally unheard or silenced through environmental governance practices in protected areas. The article addresses 'hidden' contestations over land and resource use in the Boland area of South Africa's Western Cape Province. Olivier tackles the paradox of how organisations, and in particular conservation authorities in the region, deal with a group of people known as 'bossiedokters', health practitioners who are generally denied access to the protected areas they need to access. This appears to belie promises of greater openness and more local participation in decision-making around protected areas in the post-apartheid era; the broader context of land reform seems to have little impact on established conservation practices, which continue to prioritise environmental management governance regimes in protected areas.

On the other side of the country, in the KwaZulu-Natal province, Hansen's work is also intended to render audible marginalised voices drowned out by the conceptions and practices of state agencies at the Isimangaliso Wetland Park, a world heritage site. In her article, Hansen uses Lefebvre's influential ideas about the production of space to think through hidden spatialities that are not taken into account by those responsible for the park's governance. Like Olivier, she brings to the surface contestations over conservation space and the consequences of biodiversity conservation regimes for local people. Hansen also comments on the consequences of the neoliberal conservation ideology adopted by park authorities for everyday life and livelihood practices in the area.

While not specifically addressing conservation space, the article by Salverda makes a useful contribution in discussing the persistence of colonial forms of land ownership in post-colonial Africa. Questions about elite and exclusive control of space in the form of private land tenure remain uncomfortable in many former settler colonies. Salverda's subject is the persistence of unequal land distribution on the island state of Mauritius. The article highlights in particular the ambiguous position of the old Franco-Mauritian landowning elite who remain in control of their estates. In general, landowners are able to defend their position through an uneasy collaboration with the public and private sectors, who wish to safeguard the island's image as a stable state and an attractive tourist destination. However, as Salverda shows, at times the landowning elite has been confronted with its colonial past and has strategically surrendered parts of their land.

The neoliberalisation of conservation spaces

Like the articles discussed under the previous theme, Barrett's work on peace parks in southern Africa also provides new insights into the spatiality of conservation, especially the (ongoing) securitisation of border zones. She also shows how the peace park vision partakes in the processes of neoliberal nature commodification. As Barrett argues, the peace park vision is constructed around a notion of 'exception-alism', as the parks are marketed around the idea of exceptional landscapes, wildlife and people. In the process commodified identities are assigned to the landscapes, wildlife and people within the park borders, and people who do not fit into the vision become more vulnerable. Under the neoliberal conservation paradigm, the border zones remain as spaces of physical and symbolical exclusion – and not 'spaces of opportunity', as argued by peace parks advocates. The complex realities of the parks tend to be obscured by 'logical' and 'effective' market-based solutions. As in the case of the 'bossiedoktors' (Olivier 2013, this issue), there is a silencing of displaced groups whose identities and histories are overshadowed by the parks' conception and governance practices.

The neoliberalisation of nature and conservation is also a key theme in Ramutsindela and Shabangu's article on state protected areas and land reform in South Africa. The politics of land and agrarian relations are inseparable from the neoliberalisation of nature and the alignment of conservation and business interests – with significant effects on the outcome of land claims. In their analysis of the famous Makuleke land claim in the Kruger National Park, Ramutsindela and Shabangu offer a fresh perspective on why this initially celebrated example of land restitution has not been repeated in the case of other land claims on state protected areas. The specific timing and political circumstances of the settlement solution (later known as the Makuleke model) help to explain its perceived success, but neoliberal influences on tourism policies and conservation agendas, it is argued, contributed to a perception of the model as a threat to the park and the planned transfrontier conservation area.

The last paper, by Nel and Hill, traces the way local people in Uganda, East Africa, are being sidelined in their own spaces by new land-use practices around carbon sequestration. In a now familiar theme, development efforts are negotiated by the state in a top-down approach that also draws in collaborating agents. While the article focuses on the very recent phenomenon of carbon forestry in East Africa, it highlights the continuities of colonial to post-colonial histories of land grabbing, 'green grabbing' and the politics of resource extraction. This article begins to suggest some of the ways that resources of (often marginalised) people in the South are being re-imagined under the paradigm of the current ecological and economic crisis, and acquired for increasingly global commodity markets – perhaps characteristic of trends that constitute the 'new face' of conservation and land politics in Africa.

Acknowledgements

The Guest Editors would like to extend their gratitude to academics and practitioners integral to the development and success of the Old Land, New Practices conference, not least Prof. Bram Büscher of the International Institute of Social Studies, the Hague; Saliem Fakir, Head of the Living Planet Unit of WWF South Africa and Prof Fred Hendricks, Dean of Humanities at Rhodes University. Particular thanks go to Prof Maano Ramutsindela from the Department of Environmental and Geographical Science at the University of Cape Town for his rigorous engagement, support and advice before and during the conference and crucially in the development of this thematic issue. We would also like to extend our thanks to the reviewers who provided constructive critique to the authors to transform their conference papers into publishable articles. Finally we would like to thank the following individuals and institutions for their financial support and contributions to making the Old Land, New Practices conference possible: Rhodes University Research Office and the Dean of Humanities, Fred Hendricks; the Faculty of Humanities and the School of Social Sciences, University of the Witwatersrand, in particular its former Dean Prof. Tawana Kupe and Prof. Philip Bonner who holds the NRF Chair in History; and finally the Dean of the Faculty of Natural and Agricultural Sciences, Prof. Neil Heidemann, and the Department of Geography (especially Prof. Peter Holmes and Dr. Charles Barker) at the University of the Free State.

Notes on contributors

George Barrett is a lecturer and PhD candidate in the Department of Political and International Studies at Rhodes University. Her main research interests include critical security studies, environmental politics and ethics, and an international political sociology of security, specifically ways in which neoliberal market capitalism and the commodification of life are creating and recreating processes of inclusion and exclusion, security and insecurity. George is also part of the Thinking Africa research and teaching programme which seeks to promote African theory and scholarship both within the continent but also on the international stage.

Shirley Brooks is a Human Geographer whose research and teaching has focused on environmental history, critical conservation and land issues. She is currently Associate Professor at the Department of Geography and Environmental Studies at the University of the Western Cape, having worked previously at the Universities of the Free State and KwaZulu-Natal. Shirley has written extensively on the history of protected areas and her current research focuses on the socio-spatial impacts of private wildlife production or game farming.

Jenny Josefsson is a PhD candidate in the Geography Department at University of the Free State. Her research interests include tensions between conservation and development, constructions of nature and the relationships between land, identities and power. Her doctoral research explores the impacts of conversions of cattle farms to private game farms on farm dwellers, and how they respond to these changes. This research is part of a NWO-WOTRO Programme called 'Farm Dwellers, the Forgotten People? Consequences of Conversions to Private Wildlife Production in KwaZulu-Natal and the Eastern Cape'.

Nqobile Zulu is a PhD candidate in Development Studies at the University of Witwatersrand. He is also a tutor in Sociological Theory. His research interests span across the spectrum of agrarian reform and the interface it brings to rural communities and development issues, gender and health perspectives. Nqobile is presently researching and writing his PhD on the transition from pastoral to game farming and ranching in the Groot Marico district of the North West province in South Africa.

References

Adams, William M., and Martin Mulligan, eds. 2003. *Decolonizing Nature: Strategies for Conservation in a Post-Colonial Era*. London: Earthscan.

Allan, William. 1965. *The African Husbandman*. London: Oliver & Boyd.

Anderson, David, and Richard Grove, eds. 1987. *Conservation in Africa: People, Policies and Practice*. New York: Cambridge University Press.

Barrett, George. 2013. "Markets of Exceptionalism: Peace Parks in Southern Africa." *Journal of Contemporary African Studies*. doi:10.1080/02589001.2013.808876.

Bassett, Thomas J., and Donald E. Crummey, eds. 1993. *Land in African Agrarian Systems*. Wisconsin: University of Wisconsin Press.

Benjaminsen, Tor A., and Christian Lund, eds. 2012. *Securing Land Rights in Africa*. 2nd ed. London: Routledge.

Berry, Sara S. 1993. *No Condition is Permanent: The Social Dynamics of Agrarian Change in Sub-Saharan Africa*. Madison: The University of Wisconsin Press.

Borras, Saturnino M. Jr., and Jennifer C. Franco. 2012. "Global Land Grabbing and Trajectories of Agrarian Change: A Preliminary Analysis." *Journal of Agrarian Change* 12 (1): 34–59. doi:10.1111/j.1471-0366.2011.00339.x.

Borras, Saturnino M. Jr., Ruth Hall, Ian Scoones, Ben White, and Wendy Wolford. 2011. "Towards a Better Understanding of Global Land Grabbing: An Editorial Introduction." *The Journal of Peasant Studies* 38 (2): 209–216. doi:10.1080/03066150.2011.559005.

Brockington, Dan, and Jim Igoe. 2006. "Eviction for Conservation – A Global Overview." *Conservation and Society* 4 (3): 424–470. doi:10.1111/j.1467-8330.2010.00760.x.

Brockington, Dan, and Rosaleen Duffy. 2010. "Capitalism and Conservation: The Production and Reproduction of Biodiversity Conservation." *Antipode* 42 (3): 469–484. doi:10.1111/j.1467-8330.2010.00760.x.

Büscher, Bram, and Murat Arsel. 2012. "Introduction: Neoliberal Conservation, Uneven Geographical Development and the Dynamics of Contemporary Capitalism." *Tijdschrift voor Economische en Sociale Geografie* 103 (2): 129–135. doi:0.1111/j.1467-9663.2012.00712.x.

CapeNature Stewardship Programme. *Stewardship*. Accessed April 9, 2013. http://www.capenature.co.za/projects.htm?sm[p1][category]=444.

Derman, Bill, Rie Odgaard, and Espen Sjastaad, eds. 2007. *Conflicts Over Land and Water in Africa: Cameroon, Ghana, Burkina Faso, West Africa, Sudan, South Africa, Zimbabwe, Kenya, Tanzania*. West Lansing: MSU Press.

Draper, Malcom, Marja Spierenburg, and Harry Wels. 2004. "African Dreams of Cohesion: Elite Pacting and Community Development in Transfrontier Conservation Areas in Southern Africa." *Culture and Organization* 10 (4): 341–353. doi:10.1080/1475955042000313777.

Duffy, Rosaleen. 2001. "Peace Parks: The Paradox of Globalisation." *Geopolitics* 6 (2): 1–26. doi:10.1080/14650040108407715.

Duffy, Rosaleen. 2007. "Peace Parks and Global Politics: The Paradoxes and Challenges of Global Governance." In *Peace Parks: Conservation and Conflict Resolution*, edited by S. H. Ali, 55–68. London: MIT Press.

Evers, Sandra, Marja Spierenburg, and Harry Wels, eds. 2005. *Competing Jurisdictions: Settling Land Claims in Africa*. Leiden: Brill Academic.

Ezemvelo KZN Wildlife Stewardship Programme. *Stewardship*. Accessed April 9, 2013. http://www.kznwildlife.com/index.php/conservation/stewardship.html.

Fairhead, James, Melissa Leach, and Ian Scoones. 2012. "Green Grabbing: A New Appropriation of Nature?" *The Journal of Peasant Studies* 39 (2): 237–261. doi:10.1080/03066150.2012.671770.

Godfrey, Elizabeth. 2013. "Peanut Butter Salvation: The Replayed Assumptions of 'Community' – Conservation in Zambia." *Journal of Contemporary African Studies*. doi:10.1080/02589001.2013.804245.

Hall, Ruth. 2011. "Land Grabbing in Southern Africa: The Many Faces of the Investor Rush." *Review of African Political Economy* 38 (128): 193–214. doi:10.1080/03056244.2011.582753.

Igoe, Jim, Katja Neves, and Dan Brockington. 2010. "A Spectacular Eco-Tour around the Historic Bloc: Theorising the Convergence of Biodiversity Conservation and Capitalist Expansion." *Antipode* 42 (3): 513–550. doi:10.1111/j.1467-8330.2010.00761.x.

MacDonald, Kenneth I. 2010. "The Devil is in the (Bio)diversity: Private Sector 'Engagement' and the Restructuring of Biodiversity Conservation." *Antipode* 42 (3): 513–550. doi:10.1111/j.1467-8330.2010.00762.x.

Marnham, Patrick. 1980. *Fantastic Invasion: Notes on Contemporary Africa*. London: Jonathan Cape.

Nature™ Inc. *Questioning the Market Panacea in Environmental Policy and Conservation*. Accessed April 11, 2013. http://www.iss.nl/research/conferences_and_seminars/previous_iss_conferences_and_seminars/naturetm_inc_questioning_the_market_panacea_in_environmental_policy_and_conservation/.

Neumann, Roderick P. 1998. *Imposing Wilderness: Struggles over Livelihood and Nature Preservation in Africa*. Berkeley: University of California Press.

Ngubane, Mnqobi, and Shirley Brooks. 2013. "Land Beneficiaries as Game Farmers: Conservation, Land Reform, and the Invention of the 'Community Game Farm' in KwaZulu-Natal." *Journal of Contemporary African Studies*. doi:10.1080/02589001.2013.811790.

Olivier, Lennox. 2013. "Bossiedokters and the Challenges of Nature Co-management in the Boland Area." *Journal of Contemporary African Studies*. doi:10.1080/02589001.2013.804246.

Palmer, Robin, and Neil Parsons, eds. 1977. *The Roots of Rural Poverty in Central and Southern Africa*. Berkeley: University of California Press.

Ramutsindela, Maano. 2007. *Transfrontier Conservation in Africa: At the Confluence of Capital, Politics and Nature*. Wallingford: Cabi.

Ramutsindela, Maano, and Medupi Shabangu. 2013. "Conditioned by Neoliberalism: A Reassessment of Land Claim Resolutions in Kruger National Park." *Journal of Contemporary African Studies*. doi:10.1080/02589001.2013.811791.

Snijders, Dhoya. 2012. "Wild Property and its Boundaries – on Wildlife Policy and Rural Consequences in South Africa." *The Journal of Peasant Studies* 39 (2): 503–520. doi:10.1080/03066150.2012.667406.

Zerner, Charles, ed. 2000. *People, Plants and Justice: The Politics of Nature Conservation*. New York: Columbia University Press.

Zerner, Charles, ed. 2012. *People, Plants and Justice: The Politics of Nature Conservation*. 2nd ed. New York: Columbia University Press.

Diversity behind constructed unity: the resettlement process of the !Xun and Khwe communities in South Africa

Thijs Nicolaas den Hertog

Athena Institute for Research on Innovation and Communication in Health and Life Sciences, Faculty of Earth and Life Sciences, VU University, Amsterdam, The Netherlands

The identity politics in the land distribution arrangement of the !Xun and Khwe were heavily dependent on the notion of one commonly shared community identity. However, this politically constructed identity does not match differences experienced on the ground. The !Xun and Khwe were resettled in 2004, moving from their 'temporary' settlement at an army base to a township near Kimberley. To date, they do not seem to resemble a coherent community pursuing the goal of 'cooperative production', deemed so important by land reform policies. This paper argues that forced togetherness of the past, collective identities ascribed by others and actively taken up by the !Xun and Khwe, and the socio-political context at the time of resettlement negotiations informed the delineation of community boundaries that preferred constructed unity over experienced diversity.

By late 2003 and early 2004, approximately 4500 !Xun[1] and Khwe were relocated from their 'temporary' tented camp in Schmidtsdrift, where they had lived for 13 years, to their newly developed township near Kimberley, Northern Cape, South Africa. The !Xun and Khwe came to Schmidtsdrift in 1990 when they left Namibia with the South African Defence Force (SADF) (SASI n.d.). Some !Xun and Khwe served in the SADF and were actively involved in the armed struggle against Namibian independence, the so-called border war and counter-insurgency war (Kamongo and Bezuindenhout 2011). The first signs of possible relocation from Schmidtsdrift came in 1992 when the Batlhaping people filed a land claim[2] on land which at that point was governmental property and was used by the SADF (Douglas 1997). The Batlhaping claimed land rights under the land restitution arrangement, arguing that they had been wrongfully dispossessed of their land in the 1960s (Douglas 1997; Sharp and Douglas 1996). The approval of the land claim in 1994 meant the imminent forced relocation for the !Xun and Khwe. This made the !Xun and Khwe land redistribution and resettlement arrangement somewhat different from regular restitution or redistribution arrangements in which claimants move willingly onto their acquired land or are able to acquire the land they are already living on or using for production purposes. The so-called willing seller, willing buyer policy implies voluntary resettlement on both sides. However, in this case, the restitution land claim won by the Bathlaping people displaced the !Xun and Khwe. The negotiation for resettlement conditions of the !Xun and Khwe passed several phases in which one of the important issues was how the !Xun and Khwe would be positioned in relation to each other. According to a community member and a facilitator[3] from outside the community, who took part in the negotiations, several options were on the table: the acquisition of different properties for both groups, two separate townships on one property and one township in which the !Xun and Khwe would be separated by shared facilities. Despite the strong wish and attempts of the !Xun and Khwe to go their separate ways during their resettlement, they were finally resettled together in one township.

In the Platfontein township, the !Xun and Khwe currently live together and share facilities such as a primary school and clinic. At the same time, the !Xun youth play sports within !Xun teams, the Khwe youth play in Khwe teams, tourism projects often involve only !Xun or only Khwe, and both the !Xun and Khwe have separate leadership structures. The !Xun and Khwe live together but appear to go their separate ways. This is most apparent in fact that the !Xun and Khwe reside on different sides of the township. In addition to geographical separation, their languages are also distinct (Saugestad 2004a) and form a language barrier[4]: !Xun and Khwedam languages are spoken by 56% and 35% of the Platfontein population, respectively (Letsoalo 2010). Community members have also reported outbursts of violence between the two groups as recently as 2004/2005. Separation and signs of antagonism are not the sole characteristics of the !Xun and Khwe relationship. For example, there are known cases of intermarriage between both groups and when necessary, the !Xun and Khwe are able to work together – for example, at the local radio station. The !Xun and Khwe are also united in a Communal Property Association (CPA) in which the separate leadership structures meet to discuss issues that concern both groups and to communicate and negotiate with stakeholders from outside the township, such as the provincial government and non-governmental organisations. In addition, when venturing into tourism, they portray a united image of themselves as San or 'Bushmen'.[5] Despite this united image in tourism, they have separate groups in arts and crafts projects, have built two separate cultural villages for tourism and perform traditional dances separately.

Land reform and development are often considered to go hand in hand. The redistribution grant given to the !Xun and Khwe came with the condition of a 'comprehensive resettlement and "development" scheme' (Douglas 1997, 48), which included an economic communal development plan. This is consistent with Kepe's (1999) idea of a hidden presumption of cooperative production (namely, groups of people or communities who form a legal entity work together in agricultural or other production activities) in market-led land reform, specifically the application/grant-based approach. As requested by the government, an economic development plan was commissioned by the !Xun and Khwe Trust in 1996. Several scenarios for economic development were devised, such as a diversity of farming projects, tourism, diamond mining and commercial development (SASI 2010). It is unclear how decisions were made regarding the actual development of these scenarios. However, no large-scale farming projects were visible during my fieldwork,[6] and tourism development was only present in arts and crafts projects and the Wildebeestkuil Rock Art centre. Other joint tourism developments, such as a lodge, game farming and cultural villages, had been initiated but had failed. It appears that cooperative production is a difficult goal to reach for the !Xun and Khwe and that a coherent, united community has not yet been created. The 2010 Platfontein community development plan is evident of this as it mentions the growing tension and the increasing conflicts between the !Xun and Khwe (SASI 2010). The title of the development plan, Pangakokka, translates into teamwork/cooperation (SASI 2010) which further stresses the perceived importance of overcoming tensions and conflicts for development.

Considering their explicit wish to go their separate ways during the resettlement, their sense of belonging to two distinct communities and their separate way of living in the Platfontein township, one wonders whether the apparent difficulty of 'cooperative production' could not have been foreseen. Resettlement from the military base in Schmidtsdrift was an opportunity for the communities to go their separate ways. Instead, the outcome of the resettlement negotiations delineated community boundaries in such a way that both groups were seen as one.

In this article, I will explore the identity politics and contextual factors that may have informed decisions regarding the delineation of community boundaries. First, the idea of 'natural' communities will be reviewed through a historical analysis. Second, the indigenous identity of the !Xun and Khwe and the international indigenous peoples' debate are discussed to increase our understanding of how an indigenous identity was taken up or ascribed, and recognised, during the resettlement process. And third, as the resettlement and its preceding negotiations took place in the period between 1990 and 2004, the socio-political context of moving from a racially segregated towards a unified democratic South Africa is taken into consideration. Before going further, I briefly reflect on experiences of delineating community boundaries in land reform processes.

Delineating community boundaries in land reform

Delineating community boundaries is a necessary part of land reform policies, first, because the beneficiaries must be defined. Even though South African land policy mentions both individuals and groups of people or communities as possible applicants (Government of South Africa 1997), Kepe argues that the underlying assumption is 'that coherent rural *communities* are the main beneficiary' (1999, 417 emphasis added). The origin of this focus on communities may lie in the long history of communal tenure in (South) Africa. The disruption of these systems during colonisation and the Apartheid era now presents challenges to restoring these systems within current legal structures (Cousins and Claassens 2005). Group-based rights, vested in a collection of people or communities, create the opportunity for a group of people to initiate a land claim and, when acquired, hold and manage the property according to the wishes of the community members. The Communal Property Associations Act facilitates this process in that it 'enables communities to form juristic persons' (Government of South Africa 1996, 1). The focus on communities as beneficiaries may also be illustrated by the grant-based approach in redistribution arrangements that allow households to obtain a maximum grant of R16,000 for the purchase of land, infrastructure and development (Cliffe 2000). Conventionally, these household grants are pooled to make joint projects possible, and this requires households to be grouped together to form a community. Second, delineating community boundaries is necessary as development through cooperative production is considered an indispensable step following procurement of land rights. Kepe (1999) argues that cooperative production is a hidden presumption in land reform. Groups of people or communities are apparently considered to be useful or 'natural' units for claiming land rights; however, this is not unproblematic. Both Cliffe (2000) and Kepe (1999) focus on the complexity of delineating community boundaries in land reform.

The complexity of the notions of 'community' or 'local community' is not only a challenge to South African land reform but also to other land-related approaches, such as community-based natural resource management (CBNRM). Kumar (2005), for instance, focuses on the notion of community in CBNRM, making the diversity of hidden meanings visible. Some of the characteristics that are ascribed to community are shared locality; economic, social and ethnic relations and a homogenous structure and norms and values (Kumar 2005). Kepe (1999) distinguishes yet another range of characteristics for the notion of community in land reform, namely, that definitions of communities are often based on spatial units, economic units and units consisting of a web of kinship, social and cultural relations. Both Kumar (2005) and Kepe (1999) demonstrate the difficulties of using the ambiguous notion of community in land-related decision-making processes. For example, Kepe (1999) argues that the spatial unit is often used to describe communities in South African development planning. However, it is argued that there are many features of communities that transcend spatial boundaries (economic, kinship, social and cultural relations are, for example, not necessarily bound to a defined locality), making the use of definitions based on spatial features problematic. Moreover, the artificial definition of spatial units (or communities) in the past further questions the current use of spatial units to delineate community boundaries. This, Kepe (1999) argues, may complicate land reform processes by giving rise to conflicts about who belongs to the defined community in a land claim.

The notion of community and its boundaries thus appears to be highly elusive and complex in general, and problematic in its operationalisation in land reform. At the same time, it is difficult to step away from the definition entirely and solely base land reform on individual claims. Delineation of community boundaries is a dynamic process. Boundaries can, for example, be defined by spatial, economic or social relations, by a variety of actors (for example, academics, community members and governmental institutions) and may change over time. Lamont and Molnár (2002, 168) emphasise the social construction of boundaries and define symbolic boundaries as 'conceptual distinctions made by social actors to categorise objects, people, practices, and even time and space. They are tools by which individuals and groups struggle over and come to agree upon definitions of reality'. This definition of reality is a mere representation and not an absolute truth as it only serves as a tool to 'capture dynamic dimensions of social relations' (168). These dynamic dimensions appear to become static when a notion such as community is operationalised in land reform cases; dynamic dimensions are temporarily overlooked, ignoring the multiple realities on the ground. This case study sheds light on the dynamic dimensions of social relations of the !Xun and Khwe while they become static at other levels, such as in the land reform arrangement.

'Natural' communities

The notion of *natural* communities is used in this paper to describe the way in which the !Xun and Khwe are perceived as belonging to each other beyond any doubt. This 'natural' community has probably been constructed on the grounds of historical togetherness, and the labels that have been used to describe this group of people. One community member, who took part in the negotiation process, argued that they could not be separated because they have 'come out one place, it's called Omega [the army base in Namibia],' suggesting that a historical togetherness is perceived by others. The following section examines processes behind this historical togetherness and the way in which the !Xun and Khwe were considered, by themselves and others, to be similar to and different from one another. For this paper, the starting point of the historical analysis is defined as the time at which the !Xun and Khwe were involved in the Angolan War of Independence in the 1960s, the first occasion when the Angolan !Xun and Khwe started to live in close proximity.

The Angolan !Xun and Khwe lived, respectively, in the south and south-east parts of Angola when the war of independence started in the early 1960s (Sharp and Douglas 1996). In the Angolan War of Independence, people rather than territorial ground were the main objective (Brinkman 2005).[7] The Portuguese were keen to involve the 'Bushmen' on their side because they were thought to possess great knowledge of the area and exceptional tracking skills (Battistoni and Taylor 2009). Many were incorporated into the Portuguese army as auxiliaries (*Flechas*[8]) which was the first time that the !Xun and Khwe came to live in close proximity. Despite serving in the same unit, the Portuguese made distinctions between the !Xun and Khwe which resulted in different task descriptions: the !Xun were mostly assigned as guards, while the Khwe served in offence units (Sharp and Douglas 1996). The reason that the Khwe were assigned to offence units is often credited to the social inequality between them and their Bantu neighbours, who were mostly supporters of União Nacional para a Independência Total de Angola (UNITA). The Portuguese sought to capitalise on this apparent antagonism by putting the Khwe in the front line against their former neighbours (Battistoni and Taylor 2009; Sharp and Douglas 1996).

After the independence of Angola, the !Xun and Khwe were facing retribution because of their involvement in the Portuguese army, with many local people even expressing a wish for their extermination (Brinkman 2005).[9] Most of the Khwe sought refuge in neighbouring countries (Sharp and Douglas 1996)[10], while some of the !Xun, on the other hand, were able to find refuge within the *Frente Nacional para a Libertação de Angola* (FNLA) (Battistoni and Taylor 2009; Sharp and Douglas 1996). The FNLA was a non-Marxist movement with close ties to the SADF and, after the disintegration of the FNLA, the !Xun used these ties to join the SADF (Sharp and Douglas 1996). In addition, !Xun and Khwe who had found refuge in the Caprivi Strip, Namibia, were recruited[11] by the SADF using contacts with Portuguese army officials (Robbins 2007). The SADF also recruited large numbers of local Khwe people from the Caprivi Strip Area.

In the SADF, the !Xun and Khwe served together in a special 'Bushman' battalion. The SADF made the 'Bushman' battalion an attraction, putting them on display[12] (Sharp and Douglas 1996). Contradictory to the collective identity of 'Bushmen' soldiers, distinctions were made between the !Xun and Khwe. The

Angolan Khwe were of specific interest to the SADF because of their experience of serving in offensive units and because of their knowledge of south-eastern Angola. The !Xun, on the other hand, were mostly valued for maintaining the image of a 'Bushman' battalion, as they were considered to have more stereotypical physical features. Maintaining the 'Bushman' battalion as an effective unit and fulfilling to stereotypes, forced them again to live and work in close proximity (Sharp and Douglas 1996). During their time in the SADF, the !Xun and Khwe always maintained separate living quarters. Whether this was achieved through efforts of the !Xun and Khwe or forced upon them by army personnel is unclear. Robbins (n.d.) does mention that conflicts between groups forced rigid and apparently necessary segregation in living quarters. The different positions and qualities ascribed to the !Xun and Khwe by the SADF, and segregation in living quarters, probably helped to fuel antagonism between the two groups.

After the independence of Namibia in 1990, the SADF offered the 'Bushmen' soldiers the opportunity to go to South Africa. About two-thirds of one battalion and half of another battalion opted to go to South Africa (SASI n.d.), no doubt motivated by the possibility of retribution. Approximately 3000–4000 people, 500 of whom were veterans, lived in tents on the SADF army base in Schmidtsdrift from 1990 until 2004 (SASI n.d.). Their placement at the Schmidtsdrift army base put them together once again as 'Bushmen' soldiers from Namibia and Angola. Additionally, the 'Bushman' identity was further emphasised by the !Xun and Khwe when they embraced an indigenous identity (explained in more detail below) that reinforced their collective identity to the outside world. At the same time, the !Xun and Khwe saw themselves as separate communities which is evident from their separate living areas in the tented camp and the explicit use of their names in the public domain (e.g. Smith 2004). Their aspiration for separate identities can also be recognised in the naming of a Trust[13]: they could have named it the 'Schmidtsdrift Trust' or 'Schmidtsdrift San Trust' if they had wanted to maintain a common indigenous identity. Instead, they chose to name it the '!Xun and Khwe Trust', explicitly referring to the existence of two communities. The naming of other shared structures was done in a similar fashion, for example; the *Xunkhwesa* combined school and the *XK* FM community radio station.

The idea of a natural community appears to have been constructed by a variety of collective identities, ascribed by others and taken up by them. At the same time, differences between the !Xun and Khwe were evident both within the communities and to persons or institutions that were familiar with them. At times, these differences were presented to the outside world by the combined names of shared institutions and facilities. The separate representation of the !Xun and Khwe was, however, not strong enough to break their collective identity within the resettlement negotiations.

Indigenous identity and land rights

The use of an indigenous cultural identity to build a community identity in land reform arrangements is of particular interest for the !Xun and Khwe resettlement case as they are considered to belong to one of the indigenous peoples of Southern Africa: the San (Saugestad 2004b). The term indigenous is often used to describe specific groups of people, yet there is no universally accepted definition available. In addition, it is used interchangeably with 'first nations', 'native', 'aboriginal' and 'tribal' peoples. Depending on geographical area, one concept may be used more often than the other: first nations, for example, is most frequently used in Northern America, while aboriginal is most common in Australia and New Zealand. Characteristics generally attributed to indigenous peoples comprise of: being a minority with a different language, tradition and way of life; having a relationship to a specific territory which is described as special in terms of cultural/spiritual meaning; being descendants of the 'first' occupants or having a considerably long history of occupancy compared to other peoples and having experienced a long period of *de facto* self-governance.[14]

The operationalisation of indigenous identity on the basis of these characteristics is, however, highly problematic (see Bowen 2000 for further discussion). Indigenousness, namely, the nature of being indigenous, is fraught with additional difficulties in Africa. These are connected to the high degree of historical movements of many African peoples (de Bruijn, van Dijk, and Foeken 2001) which makes it difficult for one group of people to claim they were somewhere 'first'. Some governments in Southern Africa (for example, Botswana and Namibia) have consequently denied the existence of indigenous minorities and claim that all citizens of their country are indigenous, making it difficult for communities to claim rights on the basis of an indigenous identity (Hitchcock 2002; Taylor 2007). The fourth characteristic, experiencing a period of self-governance, is also problematic in portraying an indigenous African identity, given that colonisation effectively ended any self-governance of African communities. Nonetheless, African communities do take up indigenous identities in their struggle for land resources, generally emphasising distinctions in culture and a connection to the land in terms of spiritual/cultural meaning, subsistence pattern (for example, hunting-gathering) and period of occupancy.

Embracing a San indigenous identity is often seen as a strategic attempt to effectively separate themselves from other minorities and claim certain land rights (see, for example, Hitchcock 2002; Sylvain 2002). Even though indigenous land rights are mostly unrecognised in Southern Africa, legal frameworks focusing on land dispossession require claimants to provide clear community boundaries and a genealogical connection to the dispossessed community. This means that groups of people are encouraged to 'package their claims in terms of ahistorical and bounded definitions of "tribal communities"' (Robins 2000, 60). In addition, Robins notes that lawyers representing indigenous communities have found that 'stressing aboriginal and tribal status has tended to draw positive responses and interest from general public and state' (60). Indeed, in several communications of the South African government, we can observe the different position ascribed to indigenous peoples, specifically KhoiSan. The Green Paper on Land Reform, for example, uses the term 'African people', but the authors have deemed it necessary to include in brackets 'a definition which includes the San and Khoi' (Government of South Africa 2011, 2). In the State of the Nation Address of 2012, President Jacob Zuma referred explicitly to the empowerment of Khoi-San communities through the National Traditional Affairs Bill (Government of South Africa 2012). In a recent announcement of new policy on land restitution, the African National Congress (ANC) stated: 'A re-opening of land claims "specifically for the KhoiSan people" who had until 2013 to lodge land claims' (Tolsi 2012). Lastly, during the signing ceremony of the ≠Khomani San (a collection of San people in the southern parts of the Kalahari Desert in South Africa) land claim in 1999, Derek Hanekom, the then minister for Agriculture and Land Affairs, paid explicit attention to the cultural and indigenous identity of the beneficiaries (Hanekom 1999, emphasis added):

> The quest for truth has been part of the 'Khomani San's struggle. The *revivals of the language and culture* gives proof that 'Khomani San are who they claim to be: the *first people* of this country who know the truth about the natural world and the truth about our painful history.

!Xun and Khwe indigenous identity

The indigenous identity of the !Xun and Khwe has, at times, been actively pursued by the people themselves and, at other times, assigned by others. Sans' indigenous identity is highly cultural and often contains primordial undertones, which in the past positioned them at the bottom of social hierarchy.[15] This was also the case with the !Xun and Khwe, preceding and during their time in the Portuguese and South African armies. Army personnel ascribed primordial stereotypes, such as superior tracking instincts and animal traits, to 'Bushmen' soldiers.

During their time in Schmidtsdrift and, later, in Platfontein, the indigenous identity, albeit without certain previously ascribed stereotypes, was actively sought after by the !Xun and Khwe Trust (to be transformed into !Xun and Khwe CPA), and especially by the !Xun (Sharp and Douglas 1996). They did this by representing 'Bushman' qualities in the media, attending meetings of the International Working Group of Indigenous Affairs (Sharp and Douglas 1996), affiliation with the Working Group of Indigenous Minorities in Southern Africa and the SASI (n.d.). Their indigenous identity assisted them in raising funds from international donors and was, of course, far more beneficial than their alternative image of former Apartheid 'mercenaries' (Douglas 1997; Sharp and Douglas 1996). This indigenous 'Bushman' identity is also beneficial in tourism. It is actively used in cultural tourism projects, such as arts and craft making and in a project initiated by SASI called 'Footprints of the San' that 'enables the San to utilise their traditional knowledge for a commercial benefit' (SASI n.d., 29), and the Wildebeest Rock Art Centre. The presence of ancient rock art of the San peoples probably played a role in the decision-making process regarding the purchase of land for resettlement. Indigenousness is beyond doubt part of the !Xun and Khwe identity either constructed by others or actively pursued by themselves.

Douglas (1997) discusses 'Bushman' qualities and the role of the South African government in the resettlement of the !Xun and Khwe. Although there was never any official recognition of this, indigenous identity may have played a role in the decision-making process concerning resettlement. Douglas (1997) argues that if indigenous identity was, indeed, a decisive element in resettlement, it might pave the way for ethnic group-based rights which would lead to polarisation. State intervention should, instead, be endorsed in 'accordance with the rights, whatever they may be, of South Africans' (63). The problem of ethnic group-based rights, as posed by Douglas, lies in the history of Apartheid and its group-based rights. Attributing rights on the basis of ethnicity is in sharp contrast with the aim for unity and equality. It is unlikely that governmental institutions will admit that a highly cultural or ethnic identity was decisive in giving (land) rights. At several times the

South African government did, however, seem to give indigenous peoples a 'special' position. This may be explained in several ways. First, indigenous people are said to have suffered immensely under colonialism and Apartheid and would, therefore, deserve special attention in terms of reconciliation. Second, indigenous people may serve to reaffirm a shared identity as 'African' or 'South African'. Positioning a group as the *first* people of Africa paints a picture of a shared history, namely, the 'old' and 'harmonious' Africa before colonisation and Apartheid. For whatever reason, a recognised indigenous identity assists communities in gaining positive attention from public and politics. In the case of the !Xun and Khwe, their indigenous identity probably helped create a distance from their involvement in the SADF. This, in turn, facilitated their opportunities to attain grants from the government and donations from donors. However, delineating community boundaries, consciously or unconsciously, at the indigenous San or 'Bushman' level may have made their wish of living separately more difficult to achieve.

Socio-political context

The socio-political situation at the time of the !Xun and Khwe resettlement negotiations probably informed decision-making of government officials. The !Xun and Khwe tried to go their separate ways during the resettlement negotiations but were unable to do so, hampered by governmental officials who played an important large role in the resettlement process (Douglas 1997). During the negotiations, a Northern Cape Provincial government official supposedly strongly argued that segregation belonged to the past. This emphasis on a new South Africa, without segregation, fits neatly into the nation-building rhetoric of the post-Apartheid era.

Since the start of negotiations between the 'powers that were' and the 'powers to be' in 1990,[16] South Africa has struggled to find a balance between recognising diversity and building unity. On the one hand, the pluralistic character of South Africa is undeniable but, on the other hand, unity and equality for all South Africans, irrespective of race, ethnicity or religion, were considered to be an antidote to Apartheid and the threat of a civil war (see, for example, Sparks 1996; Taylor and Foster 1999). This type of nation-building that seeks to balance diversity and equality is characteristic of multicultural nationalism (Brown 2000). Nation-building in South Africa, aside from constitutional reform, is very much a public process that aims to construct a new South Africa through a variety of symbols, thereby positioning itself further from the Apartheid past. The symbols often try to combine the recognition of diversity and unity: South Africa's motto *unified in diversity*[17] is a clear example. The notion of a 'rainbow nation' is another. The recognition of the diversity of official languages and the combination of several languages in the national anthem again acknowledges diversity and unity. The attempt to balance diversity and unity is also visible in the Bill of Rights which has a strong focus on individual rights while, at the same time, protecting the rights of religious, cultural and linguistic communities (Oomen 1999).

The balance between South Africa's recognition of diversity and aim for unity is a common topic of discussion. Bornman, for example, argues that the content of the national anthem seems to emphasise 'the ideal of unity among the South African population in striving for freedom' and concludes that 'although the flag, national anthem and the Constitution all acknowledge diversity within South African Society to some extent, the main emphasis on the symbolism and wording falls on the promotion of unity' (2006, 384–385). Oomen (1999), 83) considers that: 'The ANC (...) continues to promote diversity, but only subject to the overriding objective of national unity.' These discussions are probably fuelled by the knowledge that Apartheid government used group-based rights as a form of oppression. It is, indeed, for this reason that Oomen (1999) discusses the debate about group-based rights in the run-up to the first democratic constitution and their ultimate rejection in favour of individual rights. This may also be the reason why diversity is acknowledged but often overshadowed by the ideology of national unity.

The contradictory nature of recognising diversity and building unity is played out at two different levels; several authors make distinctions between the notion of 'nation' and 'state' (Oomen 1999), 'cultural' and 'political', or 'ethnic' and 'civil' (Jones and Smith 2001). Each of these distinctions distinguish the public level of citizenship (being a member of a country with all its obligations and benefits) and a level of personal consciousness experienced through a sense of belonging which may be vested in cultural, religious, territorial or other ascribed elements of an identity.[18] Nation-building strategies may emphasise the 'cultural' and 'political' aspects in different ways. Bornman (2006) describes Jacobinistic nation-building and syncretistic nation-building as opposites; the former emphasises the *political* level of nation and has no regard for the *cultural* level that deals with experienced diversity, while the latter emphasises the *cultural* level and ascribes specific rights to cultural, ethnic or racial groupings. In some nation-building strategies, the distinction between the *cultural* and *political* level seems to become blurred, for example, when a political entity pursues the political/civic aim to provide equal rights to everyone (at the *political* level) and, at the same time, aims to force a sense of belonging and unity (at the *cultural* level) in order to further emphasise equality. This also seems to be the case for South Africa's nation-building strategy that seems to be pulling the *cultural* and *political* levels together not only by trying to achieve equality and unity at a *political* level but also by trying to establish an experienced form of unity at the *cultural* level. Chipkin, for example, argues that the South African 'national democratic revolution ... posited the *citizen* as necessarily a member of a *nation* – as a bearer, in other words, of some or other quality of population' (2007, 99 emphasis added). Numerous studies point towards the perpetuation of experienced differences through ethnic and racial identities (Gibson and Gouws 2000; Moodley and Adam 2000) with book titles such as *Do South Africans exist?* (Chipkin 2007). Moodley and Adam (2000) point out that ethno-racial consciousness, as a legacy of Apartheid, is still widely present in South Africa. They argue that the ideological aim for emotionally experienced unity is, therefore, unrealistic. Instead, loyalty to the state (unity at the *political* level) and simultaneous recognition of diversity (at the *cultural* level) seems to be more realistic (Moodley and Adam 2000).

Recognition at a *cultural* level should, however, have consequences at a *political* level. A government that recognises experienced diversity cannot merely do so in a symbolic way. Its actions should follow this recognition, for example, in the recognition of traditional leadership structures. This results in a constant search for a balance between diversity and unity and equality. On the one hand, the South African government seems to pursue the idea that everybody is equal and similar ('we are all South Africans') but, at the same time, it recognises cultural diversity. This balance becomes especially difficult in decision-making processes where recognition of diversity or aims for equality and unity become highly visible. For example, the much-needed redistribution of wealth to battle the racially skewed socio-economic outcomes of Apartheid has increasingly become a strong focus of governmental policy land reform arrangements, a hot topic in redistribution, which are built on definitions of *groups of people* or *communities* and thereby call for a definition of boundaries. Delineating community boundaries is a process in which recognition of diversity or aims for unity become visible. Below, I will elaborate on how the delineation of community boundaries in the !Xun and Khwe resettlement seemed to be more informed by unity thinking than recognition of diversity.

Unity and diversity in land reform arrangements

The relocation of the !Xun and Khwe from Schmidtsdrift to Platfontein and the delineation of community boundaries should be understood within this socio-political context. The diversity and unity discussion at the political level has probably had an influence on the practice of land reform. In land reform policies, we can observe the previously discussed struggle to find a balance between recognising diversity and the aim for unity. On the one hand, these policies are aimed at equality with a specific focus on 'de-racialisation', hinting at uniformity by reforming racially skewed land distribution. On the other hand, these policies make use of the *categories of people* that were devised during the Apartheid era because they aim to advance those were disadvantaged in the past. This is illustrated by the fact that the main point of reference for land restitution is the 1913 Native Land Act that based land rights on racial identities which resulted in a highly unequal distribution of land to the disadvantage of non-whites (De Wet 1997). In addition, the Restitution of Land Rights Act 22 of 1994 states that legislative measures are 'designed to promote the protection and advancement of persons, groups or categories of persons disadvantaged by unfair discrimination, in order to promote their full and equal enjoyment of rights in land' (Government of South Africa 1994, 1). In land reform policy documents (e.g. the White and Green Paper on Land Reform), the group defined as 'Black people' are most often mentioned as beneficiaries of land reform policies. It is, however, unclear whether this category actually means 'non-whites' (including former Apartheid categories 'Coloured' and 'Indian'), or whether it refers to the former Apartheid category of 'Black people.' Simultaneously, the 'African' or 'South African' identities are used in a general sense to describe the citizens of South Africa, consistent with unity in nation-building. The resettlement of the !Xun and Khwe is unique in the sense that they did not live in South Africa for the greater part of the Apartheid era and thus did not experience the Native Land Act or Group Areas Act of 1950 at first hand. They were, however, involved with the SADF and its Apartheid ideologies (e.g. the creation of ethnic units). Thus, it could be argued that they were, according to the definitions in the policy documents, not necessarily a prime target for land redistribution. However, their relationship with the SADF, and the land restitution land claim by the Bathlaping people, made them part of a land redistribution arrangement as they would be displaced when the Bathlaping people won back their land.

In the resettlement negotiations, the !Xun and Khwe actively tried to go their separate ways, based on experienced cultural and social identities, but were unsuccessful. In other land reform arrangements, groups of people were encouraged to define their community through social and cultural identities. The development of a relationship between land and social and ethnic identities (Evers, Spierenburg, and Wels 2005, 3) thereby seems to be prolonged. This relationship is most prominent in land restitution arrangements. In these types of arrangements, the spatial aspect plays an important role, related to a past situation in which a group of people lived on or possessed a specific tract of land. The claimants have to prove that they used to possess that specific tract of land and that they were unrightfully dispossessed. Where they live at the time of the land claim is not important. Portraying an image of formerly coherent communities (at the time before dispossession) is likely to assist claimants in proving their previous communal residence or land possession, as it is easier to prove the past locality of a community than determining the past locality of each individual. For claimants to position themselves as a community without being able to depend on spatial features (because they do not necessarily share locality at the time of the land claim), they often depend on relational and cultural features to represent their former community identity. For the !Xun and Khwe redistribution arrangement, it was not necessary to delineate community boundaries based on social or cultural identities because it was primarily compensation for their forced removal. Therefore, community boundaries were first delineated based on the fact that they were at that point living on governmental property that would be returned to the rightful owners. The aforementioned role of the indigenous identity also seemed to have played a role during the organisation of the land reform arrangement, however, using this identity would have resulted in similar community boundaries. Similar to the sensitivity of recognising indigenous rights, the recognition of ethnic and cultural identities in land reform arrangements seems to be highly sensitive and 'unity' seems to be informing decisions rather than recognition of diversity. The nation-building rhetoric of South Africa was at that time highly focused on unity, accompanied by a heightened sensitivity for ethnic and racially based rights. Segregation based on cultural and ethnic distinctiveness, as proposed by the !Xun and Khwe, would have been highly controversial in the post-Apartheid area. For the new government, this would have come too close to the idea of ethnic-based rights and separate development. The pragmatic reasons put forward against separation, such as the required township structure and the additional costs of constructing two separate townships, no doubt also played a role. However, in my opinion, this might have been avoided if, initially, two smaller tracts of land had been purchased instead of three very extensive farms. Although this might have been financially feasible, it might not have been politically feasible for the reasons outlined above.

Concluding remarks

The dynamics of the community boundary delineation of the !Xun and Khwe suggest that there were several interrelated events and identity politics that led to a delineation that combined the !Xun and Khwe in a single space. First, their forced togetherness from their time in Angola and Namibia and their collective identities as 'Bushmen' (soldiers), San, and indigenous peoples, ascribed by others and taken up by the !Xun and Khwe, made it difficult to position themselves as two different communities. Second, the indigenous identity played out at the political level and recognised in policy documents and political statements reinforced their collective identity. Third, the socio-political context at the time of resettlement negotiations was such that recognition of ethnic or cultural diversity and segregation in land-related issues was highly sensitive. This left little opportunity for the !Xun and Khwe to go their separate ways during resettlement.

The land redistribution arrangement of the !Xun and Khwe illustrate old ways of delineating community boundaries in the supposedly new era of South Africa. The pursuit of unity seems to stand in the way of recognition of diversity and stepping away from predefined categories of people in land reform, whether it concerns categories such as 'Black people', 'Indian', 'Coloured' or 'San/Bushman'. In a similar manner as Apartheid categories, great diversity is hidden in the 'San/Bushman' category which has been forgotten or ignored during the decision-making process. These circumstances led to a situation of forced integration that may be just as harmful as forced segregation. Even though each land reform arrangement has its own unique characteristics, the insights from this case study may be informative for understanding the stagnant development in other 'communities'.[19]

Acknowledgements

For their comments and helpful suggestions on earlier drafts and intriguing questions the author wishes to thank the anonymous *JCAS* reviewers, Julie Grant, Lungisile Ntsebeza, Stasja Koot, Saskia Welschen, Harry Wels, Keyan Tomaselli and Sarah Cummings.

Notes

1. Symbols such as '!' and '≠' are used to indicate 'click' sounds in pronunciation.
2. Before the official dawn of the democratic South Africa in 1994, the national government sought to redress certain wrongs of the Apartheid past and created the Advisory Commission on Land Allocation to redress land issues (Douglas 1997).
3. Interviews took place in March and April of 2012 and were done on a confidential basis, therefore, all information derived from interviews is anonymous.
4. Communication between the two groups is facilitated by translations by people who speak both languages and by using Afrikaans as a lingua franca (spoken by 43%) (Letsoalo 2010).
5. The naming of San communities is under continuous negotiation; names such as 'Bushmen' are often said to have derogatory connotations. However, in other occasions, these names are actively taken up by San peoples. 'San' is often considered to be the politically 'correct' name, although at times also perceived as having derogatory connotations.
6. Fieldwork took place in March and April of 2012. This study is part of a Ph.D. project that is funded by the Athena Institute of the VU University, Amsterdam.
7. Brinkman (2005) is one of the rare English sources to describe events during the Angolan War of Independence through stories of refugees. It must be said that refugees and army officials are the only ones heard; the !Xun and Khwe of the Flechas unit were not heard. Portuguese sources may yield more details concerning the involvement of the !Xun and Khwe in the Portuguese army.
8. 'Flechas' literally translates into 'arrows' (Brinkman 2005) but is also translated as 'irregulars' and those whose members came from different places (like arrows).
9. In the end, thousands of Angolan San were killed in the periods just before and after independence (Battistoni and Taylor 2009; SASI n.d.).
10. The offensive role of the Khwe left them with no other choice than to find refuge in neighbouring countries (Sharp and Douglas 1996).
11. Incentives for joining the SADF are diverse, from economic to social benefits, at the same time fear for retribution of Angolan liberation forces could also be seen as an incentive (Battistoni and Taylor 2009).

12. 'Bushmen' have a long history of being put on display, see for example Gordon (1992) and Skotnes (1996).
13. A trust was formed to 'address the needs and rights of the !Xun and the Khwe' (SASI n.d., 24) and could also be seen to be as a response to the uncertain future due to the changes in the government of South Africa and also the transformation of the SADF into the South African National Defence Force.
14. These characteristics are visible in the United Nations' Declaration on the Rights of Indigenous Peoples (United Nations 2007).
15. For a broader contextualisation of the history of the San or 'Bushmen' please refer to Gordon (1992), le Roux and White (2004), Wilmsen (1989), and Wilmsen and Denbow (1990).
16. The release of Nelson Mandela and the unbanning of liberation organisations made 1990 an important year in the preparation for democracy.
17. On its website the ANC positions this phrase as a key element in their origin, political struggle and current vision.
18. Jones and Smith (2001) use the terms 'ascribed' and 'voluntary' for sense of belonging and citizenship respectively.
19. Cousins and Claassens (2005, 35) mention, for example, the dysfunctional nature of many CPAs and community land trusts and the idea that '(m)embers have often retained ties to their original communities, rather than seeing themselves as belonging to the new social entity'.

Note on contributor

Thijs N. den Hertog is a PhD student at the Athena Institute for Research on Innovation and Communication in Health and Life Sciences, Faculty of Earth and Life Sciences, VU University, the Netherlands. The PhD. study focuses on mental health processes within the !Xun and Khwe communities. He can be contacted at: t.n.den.hertog@vu.nl

References

Battistoni, Alyssa K., and Julie J. Taylor. 2009. "Indigenous Identities and Military Frontiers: Reflections on San and the Military in Namibia and Angola, 1960–2000." *Lusotopie* 16 (1): 113–131. doi:10.1163/176830809788553156.

Bornman, Elirea. 2006. "National Symbols and Nation-Building in the Post-Apartheid South Africa." *International Journal of Intercultural Relations* 30 (3): 383–399. doi:10.1016/j.ij intrel.2005.09.005.

Bowen, John R. 2000. "Should We have a Universal Concept of 'Indigenous Peoples' Rights'? Ethnicity and Essentialism in the Twenty-First Century." *Anthropology Today* 16 (4): 12–16. doi:10.1111/1467-8322.00037.

Brinkman, Inge. 2005. *A War for People: Civilians, Mobility, and Legitimacy in South-East Angola During MPLA's War for Independence.* Köln: Köppe.

Brown, David. 2000. *Contemporary Nationalism: Civic, Ethnocultural, and Multicultural Politics.* London: Routledge.

Chipkin, Ivor. 2007. *Do South Africans Exist? Nationalism, Democracy, and the Identity of the People.* Johannesburg: Wits University Press.

Cliffe, Lionel. 2000. "Land Reform in South Africa." *Review of African Political Economy* 27 (84): 273–286. doi:10.1080/03056240008704459.

Cousins, Ben, and Aninka Claassens. 2005. "Communal Tenure 'from above' and 'from below': Land Rights, Authority and Livelihoods in Rural South Africa." In *Competing Jurisdictions: Settling Land Claims in Africa*, edited by Sandra J. T. M. Evers, Marja J. Spierenburg, and Harry Wels, 21–54. Leiden: Brill.

de Bruijn, Mirjam, Rijk van Dijk, and Dick Foeken. 2001. *Mobile Africa: Changing Patterns of Movement in Africa and Beyond.* Leiden: Brill.

De Wet, Chris. 1997. "Land Reform in South Africa: A Vehicle for Justice and Reconciliation, or a Source of Further Inequality and Conflict?" *Development Southern Africa* 14: 355–362. doi:10.1080/03768359708439970.

Douglas, Stuart. 1997. "Reflections on State Intervention and the Schmidtsdrift Bushmen." *Journal of Contemporary African Studies* 15 (1): 45–66. doi:10.1080/02589009708729602.

Evers, Sandra J. T. M., Marja J. Spierenburg, and Harry Wels. 2005. "Introduction Competing Jurisdictions: Settling Land Claims in Africa, Including Madagascar." In *Competing Jurisdictions: Settling Land Claims in Africa*, edited by Sandra J. T. M. Evers, Marja J. Spierenburg, and Harry Wels, 1–19. Leiden: Brill.

Gibson, James L., and Amanda Gouws. 2000. "Social Identities and Political Intolerance: Linkages within the South African Mass Public." *American Journal of Political Science* 44 (2): 278–292. doi:10.2307/2669310.

Gordon, Robert J. 1992. *The Bushman Myth: The Making of a Namibian Underclass*. Boulder: Westview.

Government of South Africa. 1994. *Restitution of Land Rights Act 22 of 1994*. Pretoria: Department of Land Affairs.

Government of South Africa. 1996. *Communal Property Associations Act 28 of 1996*. Pretoria: Department of Land Affairs.

Government of South Africa. 1997. *White Paper on South African Land Policy*. Pretoria: Department of Land Affairs.

Government of South Africa. 2011. *Green Paper on Land Reform, 2011*. Pretoria: Department of Rural Development and Land Reform.

Government of South Africa. 2012. *State of the Nation Address by his Excellency Jacob G. Zuma*. President of the Republic of South Africa on the Occasion of the Joint Sitting of Parliament, Cape Town.

Hanekom, Derek. 1999. "Speech by Derek Hanekom, Minister for Agriculture & Land Affairs, on the Occasion of the Signing of Land Claim Agreements with the 'Khomani San and Mier Communities'". The speech was given at the signing ceremony organized by the Department for Agriculture and Land Affairs, Askham.

Hitchcock, Robert K. 2002. "'We Are the First People': Land, Natural Resources and Identity in the Central Kalahari, Botswana." *Journal of Southern African Studies* 28 (4): 797–824. doi:10.1080/0305707022000043520.

Jones, Frank L., and Philip Smith. 2001. "Diversity and Commonality in National Identities: An Exploratory Analysis of Cross-National Patterns." *Journal of Sociology* 37 (1): 45–63. doi:10.1177/144078301128756193.

Kamongo, Sisingi, and Leon Bezuidenhout. 2011. *Shadows in the Sand: A Koevoet Tracker's Story of an Insurgency War*. Pinetown: 30 South.

Kepe, Thembela. 1999. "The Problem of Defining 'Community': Challenges for the Land Reform Programme in Rural South Africa." *Development Southern Africa* 16 (3): 415–433. doi:10.1080/03768359908440089.

Kumar, Chetan. 2005. "Revisiting 'Community' in Community-Based Natural Resource Management." *Community Development Journal* 40 (3): 275–285. doi:10.1093/cdj/bsi036.

Lamont, Michèle, and Virág Molnár. 2002. "The Study of Boundaries in the Social Sciences." *Annual Review of Sociology* 28 (1): 167–195. doi:10.1146/annurev.soc.28.110601.141107.

le Roux, Willemien, and Alison White. 2004. *Voices of the San: Living in Southern Africa Today*. Cape Town: Kwela.

Letsoalo, Thabo. 2010. *San Study Baseline Report 2010*. Durban: Aids Foundation South Africa (AFSA).

Moodley, Kogila, and Heribert Adam. 2000. "Race and Nation in Post-Apartheid South Africa." *Current Sociology* 48 (3): 51–69. doi:10.1177/0011392100048003005.

Oomen, Barbara. 1999. "Group Rights in Post-Apartheid South Africa-the Case of the Traditional Leaders." *Journal of Legal Pluralism and Unofficial Law* 44: 73–103.

Robins, Steven. 2000. "Land Struggles and the Politics and Ethics of Representing 'Bushman' History and Identity." *Kronos: Journal of Cape History* 26: 56–75.

Robbins, David. (n.d.). *A San Journey: The Story of the !Xun and Khwe of Platfontein*. Kimberley: Sol Plaatje Educational Trust.

Robbins, David. 2007. *On the Bridge of Goodbye: The Story of South Africa's Discarded San Soldiers*. Johannesburg: Jonathan Ball.

Saugestad, Sidsel. 2004a. "Khoe-San Languages, an Overview." In *Indigenous Peoples' Rights in Southern Africa*, edited by Robert Hitchcock and Diana Vinding, 250–252. Copenhagen: IWGIA.

Saugestad, Sidsel. 2004b. "The Indigenous Peoples of Southern Africa: An Overview." In *Indigenous Peoples' Rights in Southern Africa*, edited by Hitchcock Robert and Diana Vinding, 22–41. Copenhagen: IWGIA.

Sharp, John, and Stuart Douglas. 1996. "Prisoners of Their Reputation? The Veterans of the "Bushman" Battalions in South Africa." In *Miscast: Negotiating the Presence of the Bushmen*, edited by Pippa Skotnes, 323–329. Cape Town: University of Cape Town Press.

Skotnes, Pippa. 1996. *Miscast: Negotiating the Presence of the Bushmen*. Cape Town: University of Cape Town Press.

Smith, M. 2004. "Long Wait for !Xun and!Khwe Homecoming Ends." *Independent* [Online], January 20. http://www.iol.co.za/news/south-africa/long-wait-for-xun-and-khwe-homecoming-ends-1.121037

South African San Institute (SASI). (n.d.). *A 14-Year Review on SASI and the San in South Africa 1996–2000*. Kimberley: SASI.

South African San Institute (SASI). 2010. *Pangakokka Platfontein Community Development Plan*. Kimberley: SASI.

Sparks, Allister H. 1996. *Tomorrow is Another Country: The Inside Story of South Africa's Road to Change*. Chicago: University of Chicago Press.

Sylvain, Renée. 2002. "'Land, Water, and Truth': San Identity and Global Indigenism." *American Anthropologist* 104 (4): 1074–1085. doi:10.1525/aa.2002.104.4.1074.

Taylor, Julie J. 2007. "Celebrating San Victory Too Soon?" *Anthropology Today* 23 (5): 3–5. doi:10.1111/j.1467-8322.2007.00534.x.

Taylor, Rupert, and Don Foster. 1999. "Advancing Non-Racialism in Post-Apartheid South Africa. In *National Identity and Democracy in Africa*, edited by Mai Palmberg, 328–341. Uppsala, Sweden: Nordic Africa Institute.

Tolsi, N. 2012. "ANC Introduces New Policy on Land Restitution." *Mail and Guardian* [Online], June 30. http://mg.co.za/article/2012-06-30-anc-introduces-its-new-policy-on-land-reform

United Nations. 2007. *United Nations Declaration on the Rights of Indigenous Peoples*. New York: United Nations.

Wilmsen, Edwin N. 1989. *Land Filled with Flies: A Political Economy of the Kalahari*. Chicago: University of Chicago Press.

Wilmsen, Edwin N., and James R. Denbow. 1990. "Paradigmatic History of San-Speaking Peoples and Current Attempts at Revision." *Current Anthropology* 31 (5): 489–524. doi:10.1086/203890.

Bossiedokters and the challenges of nature co-management in the Boland area of South Africa's Western Cape

Lennox Olivier

Department Sociology and Social Anthropology, University of Stellenbosch, Stellenbosch, South Africa

In 2007, the National People and Parks Programme was rolled out as a platform for co-management between successful land claimants, indigenous natural resource user groups and conservation authorities. It aimed to promote social 'transformation' in South African conservation management. This paper engages with the efforts made by CapeNature Conservation Board and the Boland indigenous healers – *Bossiedokters* – to resolve conflict around illegal harvesting of indigenous medicinal flora from protected areas. Dialogues emerging around such co-management platforms reveal that inequalities voiced by healers are once again silenced by government practices ostensibly designed to resolve them. Conceptualising this conflict through the lens of 'environmentality' suggests its usefulness, as well as its limitations in grasping contemporary South African dilemmas about transformation of nature. *Bossiedokters* reveal a substantially different way of being-with-nature in comparison to historically produced dominant conceptions of nature. This difference cannot be understood outside the complex relations from which they emerge and allows a better understanding of the social condition for the possibility of their voices to be heard today. While *Bossiedokters* want to reclaim their pre-colonial social authority, the question remains as to how and whether they will be able to transform conservation practice before conservation practice transforms them.

South Africa's democratic Constitution provided a framework for new environmental laws to emerge and promote social and political transformation in environmental management. The cultural practices of marginalised indigenous people were to become recognised as they were to be included in protected area planning and management.[1] In this paper, I look critically into a government agency's environmental programme and its institutional discourse on community co-management strategies. Using Agrawal's (2005) concept of *environmentality*, I

This paper derives from the author's 2012 MA thesis, entitled 'RasTafari Bushdoctors and the Challenges of Transforming Nature Conservation in the Boland Area' at the Department of Sociology and Social Anthropology at Stellenbosch University. It was presented at the 2012 'Old Land – New Practices?' Conference in Grahamstown, where it won the best student paper award. Being part of a longitudinal participatory study initiated in 2007, it will hopefully extend into a PhD study.

elaborate on the ways in which an analysis of environmental forms of govern-mentality (Foucault 1991; Luke 1999) can be paired with a historically specific approach to co-management in South Africa. Through my ethnographic account of the dialogue between CapeNature Conservation Board – managing agency of the Western Cape People and Parks Programme and the Kaapse Bossiedokters – a small group of indigenous healers and herbalists, I argue that CapeNature reflects government's limited capacity to implement co-management policy for three main reasons. Firstly, the dominant understanding of the 'fortress' character – where people and nature remain separated – of South Africa's environmentality. Secondly, the bureaucratic machinery of this particular agency, or differently put, the hierarchy of expertise which tends to limit more democratic participation. Thirdly, the manner in which unequal power relations shapes communication between 'insider' and 'outsider' and continues to privilege the former at the cost of the latter.

I begin by describing CapeNature's attempt at progressive environmentality, via their official discourse in the media along with staff interviews and the interviews and participant observation of Bossiedokters and CapeNature dialogues. These inter-views were conducted over a period of three years (2008–2011) of participatory research, during which I acted as a go-between and administrator for the Bossiedokters, keeping records at meetings and drafting various documents and reports on request. The images of participating healers are widely displayed in CapeNature reports and marketing campaigns as colourful examples of cultural diversity and community participation. However, my data suggests that the Bossiedokters remain convinced that CapeNature policies reinforce oppressive practices inherited from Apartheid. Bossiedokters believe their continued exclusion is justified through CapeNature's sustainability concerns, which demands from healers a model of scientific research to monitor and govern their engagement with nature. According to Bossiedokters, the conflict which underlies this dialogue resides not only in the laws and institutional protocol, but especially in fundamental cultural differences regarding nature and conservation. What appeared to be a battle between environmental rights at odds with human rights, takes the shape of a more complex kind of environmentality; one which silences the very social inequalities which it aims at confronting.

Apparatuses of environmentality?

> [C]onservationists are acting as gatekeepers to a discussion table that does not have a place set for those whose homeland's future hangs in the balance ... In the real world, conservation of forests and justice for biodiversity cannot be achieved until conserva-tionists incorporate other peoples into their own moral universe and share indigenous peoples' goals of justice and recognition of human rights (Alcorn 1993, 426).

As part of post-Apartheid land restitution processes, the Department of Environ-mental Affairs rolled out the People and Parks programme in 2007 to consummate the state mandated marriage between environmental rights and the rights to cultural belonging. Successful land claimants, who settled for co-management agreements with existing conservation agencies, received ownership in title even though the conservation status of protected areas remained intact (Walker 2010, 281). In the

Western Cape, CapeNature Conservation Board (hereinafter, CapeNature) manages the process. A small group of Rastafari Bossiedokters[2] became part of CapeNature's Community-Based Natural Resource Management (hereinafter, CBNRM) Programme, and were invited along with traditional healers, such as Xhosa-speaking Sangomas, to participate as a Natural Resource User Group (NRUG) of the Boland area. Living within the various Coloured communities of Stellenbosch, Franschoek and Paarl and providing their communities with medicinal and spiritual herbs and health counselling, Bossiedokters collect, trade and consume medicinal indigenous flora or Fynbos (literally, 'Fine Bush') as an integral part of their everyday practices and rituals. Due to such practices, Bossiedokters have engaged in often violent clashes with private land owners and particularly with the staff of CapeNature.

After 1994, the idea of co-management as a model for protected areas started to feature in the objectives of international CBNRM movements, linking social justice with environmental management agendas (Walker 2010). In effect, it demanded that a place should be set at the conservation discussion table for indigenous people to become an integral part of the transformation of conservation management and an alternative model to the existing forms of top-down management practised by conservation authorities (Brosius, Tsing, and Zerner 2005, 28). CBNRM is founded on the premises that local populations are greatly invested in the sustainable use of natural resources, more informed about the 'intricacies of local ecological processes and practices' and are competent to effectively manage their resources through their own forms of access (1). Since 1995, this is considered by many international conservationists to be the most effective democratic model for emancipatory enterprise in conservation management, a managerial apparatus that is able to address and contribute to the eradication of social, cultural and economic injustice.

CapeNature implemented its own version of this model after 2003. According to Cingiswa, a Sangoma who has been involved with CapeNature since the beginning of its CBNRM programme and who served for three years as the chairperson of the Western Cape People and Parks steering committee, the decision to involve communities in conservation management in South Africa through legislation resulted from the World Summit on Sustainable Development held in Johannesburg in 2002 (Cingiswa, Interview, Stellenbosch, 29 February, 2011). For her, despite other shortcomings, the Summit motivated the Department of Environmental Affairs to act upon the already common sense that there is indeed a space for communities to benefit from environmental protection programmes. This notion has become an important premise of the 'people and parks' discourse within ANC land restitution projects (Walker 2010, 280) since the outcomes of the Rio Earth Summit in 1992. In 2003, the World Parks Congress was hosted in Durban to provide a platform from which communities could speak and be heard. According to its official report, the Congress aimed at the 'incorporation of protected areas in government policy, the development of protected areas as sound business propositions and the sustainable management thereof with the involvement and support of local communities' (World Parks Congress 2003). It resolved to 'establish and implement mechanisms to address historical injustices caused through the establishment of protected areas…with special attention given…to access natural resources and sacred sites within protected areas' (Oteng-Yeboah et al. 2005, 139–218).

The recommendations called for the restitution of land, territories and resources for indigenous people whose land was declared protected areas during Apartheid.

Protected area authorities were summoned to support traditional practices and knowledge of indigenous people, and for critical review of existing conservation laws and policies. Finally, it was recommended that all members of the International Union for Conservation of Nature (IUCN) should work 'with the full participation of indigenous people, to support indigenous people' initiatives and interests regarding protected areas' and to provide support and funding to indigenous people for 'community-conserved, co-managed and indigenous-owned and managed protected areas' (Oteng-Yeboah et al. 2005, 198–199). These recommendations were also influential in the establishment of the National People and Parks Programme a few years later to ensure access of traditional practitioners to natural resources. Representatives from a number of rural communities who lived in or near protected areas in South Africa (Richtersveld, Khomani San, Riemvasmaak, Makuleke areas and iSimangaliso/St Lucia) attended a meeting at Cape Vidal on the eve of the World Parks Congress. The Director General of the Department of Environmental Affairs and Tourism facilitated follow-up meetings – known as the People and Parks Forums – to draught a plan to ensure the rights of communities affected by conservation programmes (Department of Environmental Affairs and Tourism 2004).

In 2004, the first People and Parks session was held in Swadini Forever Resort in Hoedspruit, Mpumalanga, where the focus shifted to allow all indigenous people with interests in protected areas to share *their* experiences and compare challenges faced. Most participants at this stage were land claimants. Finally, after two more conferences, one in 2006 and another in 2008, an Action Plan was developed to provide a structure and framework for the national implementation of the People and Parks Programme. The programme was shaped by the needs of land claimants who received title of ownership in several protected areas, who now had to become active in co-management arrangements. It therefore originated primarily as an attempt to find working and lasting solutions for difficulties arising from such land claims, a central part of land restitution processes. It aimed at 'empowering' and preparing the owners for full management in the future.

Meanwhile the CapeNature CBNRM Programme, together with its Local Economic Development projects, started to involve Western Cape's local communities through conservation awareness workshops. It assumed that communities needed training in conservation management prior to any 'co-management' strategies, leaving aside indigenous approaches to conservation. Although the Community-Based programme was a response to the new Environmental Act, there have been no successful land-claims in any of the conservation areas in the region so far. The community liaison committees consisted almost exclusively of Natural Resource User Group (NRUG) representatives and the status of community participants was therefore, significantly different from other provinces where participants were mostly new landowners. However, the People and Parks Programme's focus to facilitate co-management presented itself as a useful vessel by which government could uncover the environmental and social needs of all surrounding indigenous groups. In protected areas where land claims have not produced new owners (yet), the programme served as a tool to source 'authentic' community representatives. It provided a platform for interaction and motivated the formation of community Steering Committees. Steering Committee representatives were to become the voice of 'the people', channelling their needs and concerns to conservation authorities.

Local community representatives were to engage with their local reserve managers through Liaison Committees. The aim was to identify and implement projects that would encourage local community participation, allow sustainable harvesting and cultivate conservation awareness. Communities were to benefit from various economic development projects and engage actively in co-management processes. Reserve managers were given the task to facilitate the process and find resolutions for local problems by accommodating local knowledge and skills. In this manner, conservation agencies remained the gatekeepers of protected areas, and could continue to control the management process, now with the input of 'the people'. The power 'the people' would have in conflicts or disputes that emerge during their interaction remained, nonetheless, very limited. That is the case of the Bossiedokters, and the nature of their conflict with CapeNature.

Interaction between CapeNature and Bossiedokters

Until 2008, the creation of Community-Based forums in CapeNature meant a progressive first step towards 'co-management' in the Western Cape. Bossiedokters are recognised as indigenous healers, who claimed cultural and genetic kinship with the seventeenth century KhoiSan, and therefore, demand to access resources – the Cape Fynbos – upon which their practice is grounded. In the Boland area, Bossiedokters' aspirations were never concentrated on attaining ownership of protected areas but gaining unrestricted access to perform their rituals and primarily, to obtain legal permission to harvest medicinal plants. CapeNature explains the focus of their CBNRM activities to be concentrated on:

> the restoration of traditional values and systems whilst relieving pressure on natural resources; utilisation of the protected areas for neighbouring community cultural, spiritual and traditional practices; utilisation of the protected areas for sustainable harvesting of natural resources; increased community participation in the management and enjoyment of the CapeNature Protected Areas (CapeNature 2011).

The vision statement reads as '[t]he establishment of a successful "Conservation Economy" embraced by all citizens of the Western Cape, and to transform biodiversity conservation into a key component of local economic development in the province' (CapeNature 2011).

CapeNature's policy stipulates that individuals can only obtain special access to restricted areas by becoming members of an organisation, which negotiates access on behalf of its members by signing a Memorandum of Understanding (MoU) with the local reserve manager. The Siyabulela Brochure – which translates from isiXhosa as 'we are grateful' – suggests, for example, that CapeNature's Community-based and Local Economic Development programmes have successfully implemented various community-based projects (Van Vuuren 2007). The CapeNature Annual Reports from 2007 up to 2010 show how they have created job opportunities and increased participation in conservation programmes. Significant efforts were made to market and present these projects to the public sector and various local and international stakeholders and funding organisations and to testify the success of community forums which formalised, and in some areas created new user group organisations that were previously informal and unstructured. Many traditional healers who worked outside of formal structures now joined healers' organisations as their only hope to gain access to resources.

In Paarl and in Stellenbosch, two such MoUs have been signed with the Nyahbinghi RasTafari churches since they qualified as organisations. The Memorandums helped to form a relationship between the Nyahbinghi churches, the Xhosa traditional healers, the RasTafari Herbal Councils, the independent Bossiedokters and the CapeNature park managers. It allowed registered group members to access reserves for spiritual and cultural activities, and intended to set the stage for the development of the long awaited and very much sought-after Memorandum of Agreement (MoA). This agreement would be signed with the RasTafari Herbal Councils as opposed to the Nyahbinghi churches, and would only allow Bossiedokters who were registered members of the Herbal Councils to apply for harvesting permits. The Bossiedokters, being a specialised sub-group of the local RasTafari community (Olivier 2010), claimed exclusive rights to indigenous medicinal resources, since they see themselves as responsible for sourcing, harvesting and administering herbs to their communities, according to a knowledge system reserved for initiates. They believe they are spiritually called to this practice and go through prolonged training with elder healers. For the Boland RasTafari Bossiedokters, the signing of a MoU indicated the end of prosecution for the 'poaching' of medicinal plants, what they speak of as a 'legacy of colonial oppression'. It seemed to mark the beginning of formal recognition of their medicinal and spiritual practices, with the promise of a permission to harvest to follow. Since there has been no official demand from government agencies that the Bossiedokters' practices are in need of revision, Bossiedokters believed that CapeNature would become the platform from which to make claims to government authorities.

However, the administrators of the church would have to comply with certain rules, such as informing CapeNature at least seven days prior to a visitation, and providing exact numbers of entries and locations. This arrangement did not suit the Bossiedokters at all, as they claim to have dreams or visions during rituals after which they are often instantly inspired to go to the mountain. It is only when they reach the mountain that they believe they are 'guided' where to go next. Testing the viability of the MoU, the Stellenbosch Nyahbinghi church applied to host their annual 'seven-day' Nyahbinghi ritual in Stellenbosch Jonkershoek mountain reserve. It entailed a small fire to be kept burning for seven days. According to participants, they would stay around the fire and engage in traditional drumming, singing of traditional 'chants', praying and fasting. The participating healers would collect herbs from surrounding mountain slopes for preparation of cleansing and purifying teas. The reserve manager reiterated the reserve rules, stipulating that no fires are allowed inside the park,[3] no one is allowed to stay overnight and under no circumstances would it be allowed for anyone to pick any plants. Also, the RasTafaris were not allowed to go to the nearby waterfalls for their ritualistic washing of dreadlocks, since these are popular tourist attractions. The manager wanted to avoid any 'conflict of interest'.

Using the new infrastructures provided by CapeNature for such negotiations, the issues of concern were first raised at community liaison meetings and then referred by the local manager to the senior managers at head office. An 'access' meeting was scheduled and concerns were presented and argued more extensively; access cards, fees and methods to control members were discussed. The ritualistic washing of dreadlocks was rendered by officials as a 'recreational' and not a 'spiritual or cultural' activity. According to a staff member 'the lines blurred a bit', and RasTafaris would have to pay for access, unless they can prove that their activities are in fact 'spiritual and cultural'. They expressed the concern that if CapeNature allowed one particular religious group to access the river free of charge to go 'swimming' in the pools, they could surely not refuse the next group. The RasTafaris were asked to consider whether it would be 'fair' to allow them free access to the hiking trails and waterfalls, and then refuse the Dutch Reformed Church (with its thousands of members) if they should also apply for permission for their members to go swim in the pools on Sundays. The Bossiedokters responded by stating that a historical version of the routes now sold to tourist as the 'waterfall hiking trails', had been carved by the feet of their Khoi-Khoi ancestor traditional healers when taking apprentice-healers for healer initiation rituals high up to the mountain falls. They insisted that they should be able to follow their ancestors' footsteps and to show their children the traditional paths. When no final resolutions could be obtained at this level, mostly due to the provincial project manager being absent, the RasTafaris decided to take the matter to the next provincial People and Parks meeting.

The Stellenbosch reserve manager offered an alternative river some distance from the hiking trails, where RasTafaris could go to wash their dreadlocks. However, the RasTafaris were vexed about the blatant refusal to compromise on the 'original' routes, the arguments about the legitimacy of their spiritual practices, the inflexibility of fire restrictions and the refusal to allow them to pick small amounts of herbs for immediate consumption. They were also annoyed when six months passed without any action taken to implement the formal decision made by CapeNature to issue membership cards to all Bossiedokters. The cards would allow cardholders to enter reserves without seven-day notice, but before it could be implemented, central management blocked the decision. This has caused many Bossiedokters to withdraw from further negotiations with the managers. CapeNature suggested that tourists could perhaps join or observe the RasTafari traditional rituals or receive guided tours to identify medicinal herbs that grow along the hiking trail routes, but so far, the Bossiedokters have refused. They say the sacredness of the rituals would be compromised with such interference and would undermine the purpose of the rituals. Bossiedoktors claim they would only consider such 'tours' if they were entirely separated from rituals.

The RasTafari representatives who continued with discussions recently identified an unused area in the reserve. It is secluded, a safe place to make a fire, big enough to host Nyahbinghi rituals and out of sight from main tourist routes. According to the Bossiedokters, it is not close enough to where Khoi-Khoi healers were initiated prior to the introduction of Conservation control during Apartheid, but it could serve as a suitable compromise. However, this area falls just outside of CapeNature's boundaries. It is government-owned land, under conservation status and inside the main gate manned by CapeNature staff, but it falls under the management of Mountain to Ocean (Pty) Ltd. This company leases the land from government for commercial pine (*Pina*) forestry, an invasive alien species that could be considered as one of the biggest threats to indigenous Fynbos (Le Maitre et al. 2002, 144).

Consistent with their approach to increase dialogue between communities and authorities, CapeNature arranged and hosted a meeting between the RasTafari representatives and a manager from Mountain to Ocean. The manager listened patiently to the RasTafaris request to hold rituals at the proposed area. They explained in detail and with frequent inputs from the CapeNature community manager why this space is considered suitable and safe. Without much hesitation, the Mountain to Ocean manager responded by a blatant refusal to compromise on existing policy, and would not suffer anyone to stay overnight. The area identified by the RasTafaris is situated right next to the main river, a few hundred metres away from the nearest pine forests. According to CapeNature's managers, it could, without much effort, be prepared to ensure that a single controlled fire would not be a fire-threat to surrounding vegetation. However, the manager insisted that fire laws were not negotiable. It appears as if the manager was willing to continue negotiations as long as CapeNature remained a mediating partner, but for now he insisted that it was not in his power to allow any of the proposed activities (MTO Meeting, Jonkershoek, 24 June, 2010).

From 2009 onwards, there appears to have been a decrease in emphasis on restoration of traditional values and sustainable harvesting of indigenous medicinal plants in CapeNature reports. No new community-based projects have been initiated since 2007. It appears that the ideals and aims recommended by the World Parks Congress in 2003 have only resulted in quasi-successful development of a few powerless liaison committees. NRUG and community negotiations became categorised under CapeNature's list of 'challenges, weaknesses and threats'. The CapeNature Annual reports from 2008 to 2010 reveal a large decline in the numbers of community members accessing parks for 'spiritual and cultural' purposes. RasTafari and other traditional healer activities fall under the category of 'traditional healing, patients and trainees' and numbers logged have dropped from 517 entries logged in 2009, to a mere 44 entries in 2010 (CapeNature Annual Report 2010, 19).

Exploring apologies

I started my interviews with the CapeNature reserve managers in 2009, almost a year prior to the meeting with Mountain to Ocean. I recorded and took extensive notes of their explanations of the problems they experienced over the years with RasTafari healers, and I was sympathetic towards CapeNature at the time. Everyone seemed convinced the practices of participating healers posed no serious threats to biodiversity and wanted to find a way forward. They expressed tolerance of differences, and would often bend rules and regulations to hasten the process. But there were always endless administrative demands, prerequisites and guidelines to adhere to. These demands were mostly to satisfy the so-called 'gatekeepers' – the scientific department – who justified themselves by the shared concern to protect nature from being destroyed by the practices of lay people.

Managers provided the following explanations, or what they referred to as 'issues and stumbling blocks' hampering the success of RasTafari community participation (CapeNature Reserve Manager, Interview, Paarl, 26 October, 2009). Firstly, CapeNature is in need of a champion(s) to assist them with implementation of the projects. The single Conservation Community Manager assigned to manage the whole Boland Area cannot cope on his own. In short, CapeNature is under-staffed. Secondly, the high turnover in community management staff since 2005 caused a lack of continuity. Sometimes months elapsed before a manager was replaced and resulted in previous groundwork being lost during turnovers. Thirdly, reserve staff were not trained in community management, referred to by a manager as 'an embarrassing fact' and 'a relic of the apartheid era, where reserves were managed in isolation' (CapeNature Reserve Manager, Interview, Paarl Reserve Manager, 26 October, 2009). He believes that in many cases reserve staff do not have the most basic knowledge to manage communities. Fourthly, CapeNature officials experience RasTafaris' mistrust of government authority.[4] Fifthly, the RasTafaris believe their knowledge of the veld to be sacred and it could not be shared with just anyone, particularly not with state authorities. Lastly, RasTafaris were hoping for a 'quick fix' in 2005, when the first formal community meeting in Paarl was held, and many gave up trying when they realised it was going to take many years before any harvesting would be allowed.

While I found these issues contributed to the lack of progress, I do not believe that resolving them would result in the issuing of harvesting permits. It would take far more than a couple of champions and a few years of scientific research to prove empirically that the Bossiedokter practices are indeed sustainable. The high turnover in staff sounds like a reasonable explanation if there was any significant groundwork done in the first place. There has been no turnover in community managers since 2009, but there is still no progress in the implementation of the promised new harvesting policies. Indeed, the Stellenbosch Herbal Council chairman is convinced that the last two years had made it clear that, even if community participants, community managers and reserve managers agreed on the issuing of harvesting permits, it would still not be authorised by the scientific department. An interview with the provincial community programme manager (Bellville, 28 August, 2009), revealed a significant limitation in CapeNature's capacity to implement their new harvesting policies. It was assumed as the responsibility of community representatives to engage through the CBNRM forums with their local reserve managers, and to apply for harvesting permits by following the new 'Policy on Consumptive and Commercial Utilisation of biological resources from protected areas and surrounds' – even if the document had not been distributed amongst local reserve managers.

Asked about the requirements in the policy regarding monitoring and assessment of 'the socio-economic influence on sustaining the resource sector, and sufficient measures being put in place to monitor and regularly audit the harvesting and associated impacts' (CapeNature 2007 'Draft' Policy on Consumptive Utilisation. 1. (h), (i)) the programme manager responded as follows:

> ...the only problem we have at this stage is an institutional capability to deal with harvesting of indigenous resources...because now there must be a monitoring done, who is going to do that monitoring?...we don't have staff...we have been trying to get more funding...to get more people...to be ready for that...but who will determine the harvesting levels? At this stage there is [sic] no records where people have been harvesting, there has been no information that have been recorded...we need that information, we need to understand the patterns...what is their harvesting methods? So that the scientists will need to be able to focus on those kind of things...We don't have it in our budget (CapeNature Programme Manager, August, 2009).

In short, the manager suggested it is up to the user groups to provide researchers and source funding for monitoring and impact assessments, since CapeNature do not have staff or money to do so. User groups have to come up with research proposals and harvesting proposals that include monitoring and evaluation in the design. The proposals must show that harvesting will be empirically sustainable, the exact impact it will have on the environment and how this impact will be monitored. Proposals must include a business plan with detailed discussions regarding market analysis, product and service development, marketing strategy, financial data, organisation structures and management, ownership structure, risk factors, sales profitability objectives, cash flow projections, asset acquisition schedules, projected profit and loss statements and the list goes on. For businessmen and economic enterprises, this may appear fairly standard and very 'reasonable' rational expectations; to the Bossiedokters, it appears like just another method used by the authorities to exclude them once again.

The Bossiedokters claim the Department of Environmental Affairs allocated a significant budget to CapeNature in 2007 to implement the People and Parks Programme and community projects. However, according to the programme manager there was never enough money available for cultivation projects, or for research that would allow sustainable harvesting in the reserves. The problem is conceptualised as lack of 'institutional capability', but research and empirical evidence would be sufficient to prove empirically that their harvesting is indeed sustainable. How much would such research cost, and does current harvesting impacts on biodiversity justify the cost of the research? Policies in theory allow healers to harvest, but in practice it remains impossible to do so with very little hope of obtaining permits in the near future. As CapeNature lacks the necessary funding to implement their new agreements and responsibilities to the communities, the Bossiedokters will remain seen as poachers. For them the practices of Khoi and San indigenous cultures and knowledge systems remain illegal.

Meanwhile, community and reserve managers have ensured that Bossiedokters adhere to all other requirements, leaving the most complicated issue until last. The Herbal Councils provided lists of all members with photos, copies of identity documents and other personal details. Representatives provided official letters with letterheads from their organisations stating they are indeed authentic representative members. Bossiedokters provided detailed lists of popular medicinal plants, amounts used, where and how much they intended to harvest, prices at which herbs are sold, methods of harvesting, and so on. However, when the Bossiedokters demanded an official harvesting application form, the local community manager could not produce one. His last response on the matter in 2011 was: 'I do not think that there are such forms available yet' (CapeNature Community manager, Interview, Jonkershoek, 22 September, 2011). This is four and a half years after the Consumptive Utilisation policy has been accepted for implementation. All applications are being processed verbally, and very little records are kept of discussions between community managers and community members.

Until 2009 none of the healers in Stellenbosch and Paarl had access to email. It was only after I started participating in the process that it became possible for them to keep written records of negotiations themselves. By then most of the healers had already given up hope in the CBNRM programme, but in 2010 their plan of action changed. Bossiedokters decided to create a paper trail of all negotiations. They asked me to write down all past grievances and create wish lists and recommendations to submit at the People and Parks provincial meetings. The hope was that if interactions with local managers could be properly documented, it could be channelled through the system to the relevant higher-level authorities, and eventually it would reach the Department of Environmental Affairs. They decided they could no longer trust the community management team of CapeNature. The fact that over five years of negotiations with park managers have not been able to go beyond discussion around very restricting access permits, the establishing of authenticity of their 'spiritual and cultural' activities and negotiating of issues, such as whether the making of a small fire during rituals would be allowed, created a perception amongst many healers that any further attempts to gain permits for harvesting were a waste of time. On the other hand, reserve managers claimed they received insufficient information and guidelines from their superiors, and that they are in a limited position to improvise, since they will be held responsible by the scientific department if things go wrong.

I decided to approach the 'gate-keepers' directly. The head of the scientific department proclaimed that although park managers should follow new policy guidelines where possible, they are free to allow the process to continue, even if all guidelines were not met. He suggested that the guidelines provided in the policy were not intended as laws that could prevent the implementation of any project. This contradicted information given by local and area managers, who insisted they were coerced by the guidelines to act in a particular manner.[5] Bossiedokters suggest that, their failure to successfully negotiate permission for a once-off ritual ceremony, and that their family members will still be charged tourist rates to access the waterfalls for ritualistic cleansing and washing of dreadlocks, means that no change has occurred. Furthermore, since the picking of fresh herbs for consumption in the park remains forbidden, the memorandums served no purpose. Bossiedokters suggest that the only compromise that authorities have made so far is to allow healer members to enter designated areas of the reserve under strict conditions, without having to pay daily tourist rates.[6]

At the latest signing of the MoU in 2010, the Executive Director of Operations was disappointed with the low turnout of Stellenbosch RasTafaris, assessing that they were 'not serious' about the agreement. When those who attended heard about his comments they became furious, as in their version, CapeNature informed the community of the signing event only one day in advance. They provided transport for members from Cloetesville to their Jonkershoek offices, but sent a big truck usually utilised for fire fighting operations in the reserves. RasTafari woman and men were dressed in their traditional garments making it difficult to climb up the truck's steep and slippery metal ladder. Some of the older members reported they took one look at the truck, turned around and went back home.

Why can't they send their mini-buses to come and fetch us? Now they send this old truck and expect us to climb up the side of it like baboons. These people have no respect for us, imagine that, if they had to go and fetch whites from town, would they expect them to climb up and down this truck. We are elders; we are not youths or kids. We are really tired of this same old treatment (RasTafari Elder woman, Jonkershoek, 10 June, 2010).

People and Parks

From 2008 onwards, CapeNature community management became absorbed in the implementation of Western Cape People and Parks community meetings. They have been very successful in organising quarterly meetings at various venues around the Western Cape, transporting representatives to and from meetings and accommodating them in guesthouses for the duration of the meetings. CapeNature officials facilitated the formalisation of a Western Cape Steering Committee, facilitated the democratic election of a chairperson, secretary, treasurer and so on, and went to great lengths to ensure that there are always sufficient representatives of each area present at meetings. These meetings have become a space where community representatives from different areas could interact, share experiences of struggles and identify shared needs with each other. It also provided a space where concerns could be raised and information could be exchanged about conservation practices.

Cingiswa viewed the first year of the programme especially enthusiastically. Everyone felt the involvement of Department of Environmental Affairs would speed up the process, and put more pressure on CapeNature to respond to the needs of the people. The user groups also heard rumours of a budget allocated to implement the programme, and everyone hoped the programme would be able to facilitate CapeNature to improve on the shortcomings of the Community-based projects. Their hope was rekindled for the initiation of harvesting and cultivation projects. However, after two years of quarterly People and Parks meetings, the renewed hopes were gradually smothered once more. The short reply from CapeNature to the ever-increasingly frustrated user groups remained the same; there is no money available for project implementation. Money allocated by government was apparently spent on the hosting of People and Parks Steering Committee meetings. When the chairperson and the secretary demanded that the Steering Committee should have access to a breakdown of the budget, the reply given by the provincial programme manager was that Steering Committee members would never gain access to budget information, not while he has a say in it.

In July 2010, the Steering Committee instructed their chairperson and their secretary to compile a comprehensive list of community complaints as recorded since 2008, to submit at the annual national People and Parks meeting hosted in Durban, where it would reach the ears of the Department of Environmental Affairs. The Steering Committee noticed that CapeNature officials often omitted inconvenient complaints raised at meetings from the minutes, and even when recorded in the minutes, it was simply omitted from the following meeting's agenda. Tensions increased with time, and meetings became more heated and sometimes even aggressive, and needless to say, also unproductive. The Steering Committee members felt their voices fell on deaf ears, most important issues were simply ignored, and as time went by members became more verbal about their frustrations. They started to openly accuse managers of incompetence, omitting issues raised strategically from the agendas, and silencing members who raised complaints during meetings in a disrespectful manner. They believed managers disregarded the committee's abilities to manage their own affairs. They wanted information about the spending of the People and Parks budget, and demanded clarity on the purpose of attending meetings, since it appeared CapeNature officials had no intentions to include them in any decision-making or respond to their problems, concerns and grievances. As soon as a representative would withdraw from participation, community managers would replace them with new people from different areas.

New members were often openly misinformed on why previous members were replaced. The situation came close to a breaking point towards the end of 2010, when the programme manager suspended the Steering Committee secretary prior to an investigation. He was accused of smoking cannabis in a minibus hired by

CapeNature to transport members back home after a quarterly meeting in George. The secretary initially reported the bus driver for reckless driving, speeding and talking on his cellular phone while driving. He also reported the driver for refusing to stop to allow the Bossiedokters to do their usual prayer and smoking ritual at the onset of the road trip. The driver was apparently in a hurry to another appointment and told the Bossiedokters to smoke in the bus, as long as everyone else in the bus agreed to it. When the secretary filed a complaint of reckless driving against the driver, the owner of the minibus responded with an email to the programme manager, accusing the secretary of smoking in a non-smoking bus. The CapeNature programme manager blatantly suspended the secretary for six months, accusing him of breaking the code of conduct, which states that members are not allowed to abuse drugs while participating in CapeNature activities.

After the six months suspension was announced, the local community manager and the Boland area manager were asked to start an investigation. After several interviews with relevant participants, they recommended the secretary be given a written warning not to smoke under any circumstances in public transport again, and to resume his duties as secretary without any further delay. The programme manager overruled this decision once more, reinstating the six months suspension. This is when the secretary instructed me to draft a letter to the CEO, explaining the situation. A meeting was scheduled with the CEO, who insisted that she could not lift the suspension until the programme manager submitted his report on the matter. The report was not submitted for another three months. A meeting was scheduled with the Executive Director of Biodiversity who agreed to look into the matter urgently. Several consultations were held with the legal department who confirmed the suspension as unusual and a 'previously unheard of procedure'. By now the Bossiedokters had collected a long record of emails and voice recordings of all the discussions and verbal agreements, and it took just over seven months before the six months suspension was finally revoked by the CEO. The secretary was finally restored in his position and instructions were given that the code of conduct should be revised in collaboration with the Western Cape Steering Committee members. The Bossiedokters insisted that the ritualistic use of cannabis could not be defined as drug abuse, since it forms part of what they believe to be pre-colonial Khoi and San rituals and tradition. However, most importantly, the list of complaints compiled by the secretary and the chairperson was not submitted at the National People and Parks meeting. The official who imposed the suspension nominated another representative to go to Durban, without consent of the Steering Committee. CapeNature management was apparently at the time still unaware of this list of complaints to be submitted at the National meeting. The only official who was informed about the list was, not coincidentally, the same official who implemented the suspension. The replacement community representative that was selected by the official made no mention at the national meeting of any of the complaints or the suspension of the secretary.

Many of the Steering Committee members, including the chairperson at the time, were convinced that the smoking incident provided an opportunity to the programme manager to suspend the secretary until after the national meeting, thereby preventing the list from reaching Durban. Considering the distrust and already existing frustrations, it was not surprising that this irregular suspension raised such suspicions. Meanwhile, the secretary and chairperson decided to email the list of complaints directly to the CEO, who distributed it to the relevant directors and managers. Since the list was compiled and undersigned by the chairperson from her own minutes kept since 2008, the list appeared to be her initiative. The list openly attacked and accused the senior community and programme managers responsible for People and Parks quarterly meetings, mentioning their names and accusing them of misconduct, disrespectful behaviour towards Steering Committee members, the hiding of information, the selective omitting of important inputs of community representatives and a general abuse of power.

Many of the committee members could not write or read English, could not take minutes, and could therefore not read or write CapeNature's reports and depended on the chairperson and the secretary to handle all the administrative work on their behalf. Most members had no Internet access and were not computer literate. At the last quarterly meeting in 2010, the chairperson recommended that only people who are fluent in English and who are able to perform the administrative duties expected from them, should be nominated for steering committee positions at the next Annual General Meeting. She requested for local meetings to be held prior to the annual meeting, so that more suitable candidates could be nominated, to lighten her administrative burden, since many of the members nominated by CapeNature were not competent to fulfil their tasks.

CapeNature officials accepted her recommendation. Community managers were instructed to call local meetings prior to the next Annual General Meeting, and to ensure that the re-election of suitable representatives for the next provincial People and Parks Steering Committee was prioritised. It was recommended that representatives had to be fluent in English, must be able to read and write and should preferably be computer literate. When the local community manager in the George area called a meeting, the chairperson – who lived in that area – was not invited. She was removed as community representative without any further explanations and replaced by another member. She was also not invited to the 2011 Annual General Provincial Meeting while still acting as Western Cape Provincial People and Parks chairperson, and an official member of the National People Parks executive Steering Committee. She never had the opportunity to present her annual report to the Steering Committee, or to hand over her chair position to the next 'democratically elected' chairperson, who was coincidentally the same person who was elected by the programme manager to the national meeting in Durban, while the secretary was under suspension.

> What I have seen . . . happening in Cape Nature is this. The community managers only have one aim, and that is to ensure that they organize the meetings they were told to organize annually. They make sure they get enough community representatives around the table so they can show that they have the necessary representation required to make it look legit [sic]. Then they get the communities' opinions and complaints and comments from the representatives, and they selectively minute the ones they find useful and easy to . . . address. Difficult questions are ignored, and . . . [if] you become stubborn, you are not going to be invited to the next meeting. The good ideas and easily implementable proposals are then presented by them as if it were their initiatives. It is then placed on their yearly reports as 'success stories'. Where is the voice of our people for whom the People and Parks programme was intended . . . If our aim was to take part in a collaborative and co-management interaction, now I know for sure that under these conditions the process has failed (Cingiswa Chairperson, Interview, 14 June, 2011).

Educating nature

When Bossiedokters and Sangomas speak about their experiences attending Community-based meetings, they refer to efforts made by officials to educate them on what conservation is and on compliance with CapeNature's rules and regulations. Many of the healers reported that they went to a few of the workshops and then decided to withdraw, since they experienced it as a waste of their time.

> These conservation people think we are stupid, they tell us all these things as if we are ignorant...uneducated, as if we only know how to destroy nature. Do you know how many times we told them it is not us pulling the plants out by its roots? We don't even use the barks [sic] of the trees. They want us to listen to them only, and they show us pictures of poachers and plant collectors. They even show pictures of how they harass our brethren in the market places selling herbs to the sick people. We are RasTafari, we only use plants and we know how to take care of our plants. We are ancient healers, we know from creation how we trod [walk] in nature (RasTafari Bossiedokter, Idas Valley, 2010).

The Bossiedokters believe CapeNature needs to be 'educated' about the differences amongst various traditional practices of healers, and to distinguish between healers, collectors and poachers. The healers suggest that even though collectors often sell herbs to healers, the bulk of large-scale poached herbs are sold to various small factories, where plants are processed and active ingredients are extracted for export to pharmaceutical companies. These factories are licensed to trade in medicinal plants cultivated and harvested from private land. According to the healers some turn a blind eye to the origin of the plants, since plants harvested from the wild are known to have higher levels of the sought-after active ingredients, and are in general of a higher quality than cultivated crops. Healers I interviewed denied their participation in such bulk harvesting operations. They affirm to be willing to help expose what they refer to as 'hit and run' or 'fly by night' harvesting practices and report and prevent such type of harvesting, provided that Cape Nature would allow them to harvest sustainably and legally.

Conclusion: towards a resolution to co-management conflicts

The culture of CapeNature is shaped by a long history of techno-managerial strategies, top-down implemented infrastructure and commitments to a range of financial institutions that sponsor their 'evidence-based' conservation methods. If post-apartheid conservation is serious about social change and democratic inclusion, it should not reproduce scientific prejudices about traditional practices being arbitrary, informal, and potentially destructive. This representation of traditional practice is not only reflected in conservation practices, but *produced* by it in the present: Bossiedokters can only appear as a danger to the 'nature' of political management, because they are already placed outside of it. So the fences and gates in conservation reserves are a material and symbolic mechanism that works at the same time as securing political power to CapeNature, while turning outsiders, who are excluded and who in fact suffer the violence of fence exclusion, into a 'natural' threat.

Within the first two months of my interactions with the People and Parks Steering Committee members in 2009, the chairperson and the secretary indicated a need for the traditional healers of the Western Cape to become organised under a non-governmental organisation. Bossiedokters were hoping that the People and Parks programme would support their belief that government had to clamp down on unsustainable activities on private property, to protect indigenous biodiversity by means of conservation laws and restrictions outside reserves. Bossiedokters claim that not nearly enough is done by the Ministry of Environmental Affairs to prevent private developers from destroying the indigenous environment. They believe they are being targeted, while the serious environmental threats posed for example by pine forestry and the wine industry are being ignored.

It is no surprise that healers never identified with the platforms provided by CapeNature. They desired their own independent political space. In 2010, I started to design, with key figure healers serving on the People and Parks Steering Committee, the Cape Bush Doctors (Kaapse Bossiedokters) Not-for-Profit Organisation. By August 2011, a team of 12 executive members accepted the founding constitution based on my research data. The executive consists of seven traditional healers, two academics, one lawyer and two health business owners. The main objective is formalising and protecting of indigenous healers and their medicinal knowledge in the broader Cape. The following comment of Cingiswa (Interview, 23 January, 2011), reflects on the position of healers in relation to this organisation:

> [We] take care of nature and people, for us nature and people sit together. This connection is kept alive through rituals that is done mostly in protected areas...we must be recognized and informed about the policy-changes we have been waiting for in such areas where we practice our rituals. Our ancestral lands, plants and animals awaits us...they need us as we need them, so the true healing can follow.

The healers' organisation focuses on assisting healers to become active agents of their own development and the development of the broader community. It works towards recognising and acknowledging healers and indigenous communities, and leaders who will assist their local municipalities and politicians to protect nature inside and outside reserves, with and for the people. The organisation also aims to protect indigenous healer knowledge from biopiracy, and to develop a certification programme for the standardisation of sustainable harvesting methods and invasive alien management.[7] Healers believe it is time for them to collectively confront conservation organisations that are in many ways compromising, instead of promoting their participation in environmental protection and that of their community. They believe that a unified indigenous healer organisation will become the first step towards such a project.

Not unlike Cepek's (2011) discussion of Environmental and Conservation Programmes implemented in Zabalo and other Cofan communities in the far north-east of Ecuador, the Bossiedokters appear to maintain, at least to some extent, a critical consciousness of their own practice by viewing their actions 'in terms of their political agendas and their cultural perspectives rather than the rationales of Environmental and Conservation Program agents' (505). Bossiedokters believe that the value and authenticity of their practice would be compromised if rendered a sellable product to tourists, suggesting that, they are somewhat conscious of the potential threats presented by the manner in which the market inflects culture and ethnic differences. While they are willing to sell their products, medicinal herbs and consultation to the modern world, they hope to seal off their value systems and the integrity of their practices from the market. Bossiedokters continue to see scientific intervention as something outside of their community logic and needs. They see their practices as ahead of time, believing that science will eventually confirm through research that their herbal knowledge and healing practices have 'real' medicinal value. Bossiedokters want to reclaim their authority over nature – and ultimately over their own practices – but the challenge seems to be as to how they will be able to transform conservation practices before conservation practices transform them. The battleground is ultimately one of political aspirations and should not simply be folded into the logic of environmentality.

Acknowledgements

I would like to give a special thanks to Bernard Dubbeld (promoter) for his support and guidance and especially Fernanda Pinto de Almeida for her assistance with the final editing and structuring of this paper, as well as the two anonymous reviewers for their critical responses. Many thanks to all the Bossiedokters and Sangomas who participated in this research. I hope with all my heart that this paper will somehow contribute towards their ongoing struggle for African Liberation.

Notes

1. Examples of post-Apartheid environmental legislation that allows for the recognition of cultural and spiritual practices of traditional communities are the National Environmental Management Act (1998), Biodiversity Act (2004), Protected Areas Act (2003) and National Forests Act (1998).
2. Indigenous Healers/Herbalists who refer to themselves as Bush Doctors.
3. Comaroff and Comaroff's (2001, 650) discussion of the panic and anxieties – the 'public divination' – that the 'apocalyptic' fires of January 2000 evoked, burning some 9000 hectares in the Cape Peninsula, shows how processes and discourses of nature reflects processes in society and can contribute to the depoliticisation of politics.
4. CapeNature officials often complained RasTafaris do not differentiate between them and other government officials, such as the Police Force or Military Force. They considered them all as servants of the same 'Babylon' system, the authority who enforced white supremacy under colonisation and Apartheid.
5. Hoag (2010) elaborates on how officials at the Immigration Services Branch of the South African Department of Home Affairs 'develop systems of meaning to help them mitigate the challenges posed by an unpredictable populace and management hierarchy'.
6. A daily permit to enter Jonkershoek reserve is currently R30 per person per day.
7. Biopiracy is 'the practice of commercially exploiting naturally occurring biochemical or genetic material, especially by obtaining patents that restrict its future use, while failing to pay fair compensation to the community from which it originates' (Oxford Dictionaries 2011).

Notes on contributor

Lennox Olivier completed his MA in June 2012 in Sociology at the Department of Sociology and Social Anthropology, University of Stellenbosch. He has published on the topic of South African RasTafari culture and liberation struggles in the South African Review of Sociology and Anthropology Southern Africa. He is currently involved in the formalisation of the Cape Bush Doctor/Kaapse Bossiedokter Nonprofit Organisation. His research interests are focused on the recording of the indigenous medicinal knowledge of San and Khoi elder members of the organisation, with the aim to help preserve and protect this vastly disappearing knowledge from becoming extinct. He can be contacted at: lennox@thebranch.org.za

References

Agrawal, Arun. 2005. *Environmentality: Technologies of Government and the Making of Subjects*. Durham, NC: Duke University Press.

Alcorn, Janis B. 1993. "Indigenous Peoples and Conservation." *Conservation Biology* 7 (2): 424–426. doi:10.1046/j.1523-1739.1993.07020424.x.

Brosius, J. Peter, Anna L. Tsing, and Charles Zerner, eds. 2005. *Communities and Conservation: Histories and Politics of Community-Based Natural Resource Management*. Walnut Creek, CA: Altamira Press.

CapeNature. 2007. "Protocol for the Processing of Applications to Gain Access to CapeNature Nature Reserves for Consumptive and Non-Consumptive Utilization Activities and Purposes." Draft.

CapeNature Annual Report-2007/2008. Accessed July 11, 2010. http://capenature.co.za/about. htm?sm[p1][category]=283

CapeNature Annual Report-2009/2010. Accessed July 11, 2010. http://capenature.co.za/about. htm?sm[p1][category]=283

CapeNature Annual Report-2010/2011. Accessed July 11, 2010. http://capenature.co.za/about. htm?sm[p1][category]=283

Cepek, Michael L. 2011. "Foucault in the Forest: Questioning Environmentality in Amazonia." *American Ethnologist* 38 (3): 501–515. doi:10.1111/j.1548-1425.2011.01319.x.

Comaroff, Jean, and John L. Comaroff. 2001. "Naturing the Nation: Aliens, Apocalypse and the Postcolonial State." *Journal of Southern African Studies* 27 (3): 627–651. doi:10.1080/13632430120074626.

Department of Environmental Affairs. 2004. "Deputy Minister Mabudafhasi will open a groundbreaking People and Parks Workshop." Accessed April 15, 2011. http://www.info. gov.za/speeches/2004/04102208151007.htm.

Foucault, Michel. 1991. "Governmentality." In *The Foucault Effect: Studies in Govern-mentality*, edited by Graham Burchell, Colin Gordon, and Peter Miller, 87–104. Chicago: University of Chicago Press.

Hoag, Colin. 2010. "The Magic of the Populace: An Ethnography of Illegibility in the South African Immigration Bureaucracy." *Political and Legal Anthropology Review* 33 (1): 6–25. http://onlinelibrary.wiley.com/doi/10.1111/j.1555-2934.2010.01090.x/full.

Le Maitre, David C., Brian W. van Wilgen, C. M. Gelderblom, C. Bailey, R. Arthur Chapman, and J. A. Nel. 2002. "Invasive Alien Trees and Water Resources in South Africa: Case Studies of Costs and Benefits of Management." *Forest Ecology and Management* 160 (1–3): 143–159. doi:10.1016/S0378-1127(01)00474-1.

Luke, Timothy W. 1999. "Environmentality as Green Governmantality." In *Discourses of the Environment*, edited by Eric Darier, 121–151. Oxford: Blackwell.

Olivier, Lennox. 2010. "Racial Oppression and the Political Language of RasTafari in Stellenbosch." *South African Review of Sociology* 41 (2): 23–31. doi:10.1080/21528586.2010.490378

Oteng-Yeboah, Alfred A., Nikita Lopoukhine, Paul Mafabi, and Juan M. Maldonado. 2005. *UCN Recommendations*, 139–218. Accessed July 20, 2011. http://cmsdata.iucn.org/down loads/recommendationen.pdf

Oxford Dictionaries. 2011. Accessed October 5, 2011. http://oxforddictionaries.com/definition/biopiracy

Van Vuuren, Peter. 2007. *Siyabulela – Making a Difference to People and Countryside*. Accessed July 17, 2011. http://www.capenature.org.za/docs/1209/Siyabulela%20Booklet.pdf

Walker, Cherryl. 2010. "Land Claims, Land Conservation and the Public Interest in Protected Areas." In *Development Dilemmas in Post Apartheid South Africa*, edited by Bill Freund and Harald Witt, 275–298. Scottsville: University of KwaZulu-Natal Press.

World Parks Congress. 2003. Briefing minutes. 8 April 2003. Environmental Affairs and Tourism Portfolio Committee. Chairperson: Mr Arendse (ANC). Accessed February 7, 2011. http://www.pmg.org.za/minutes/20030407-world-parks-congress-2003-briefing.

List of Interviews and Meetings

Bossiedokter, Interview, Cloetesville, 10 November 2009.

CapeNature Reserve Manager, Interview, Paarl, 26 October 2009.

CapeNature Community manager, Interview, CapeNature Conservation Offices Jonkershoek, Stellenbosch, 22 September 2011.

Cingiswa, Sangoma/People and Parks Chairperson, Telephonic Interview, 23 January 2011.

Cingiswa, Sangoma/People and Parks Chairperson, Interview, Stellenbosch, 29 February 2011.

Cingiswa, Sangoma/People and Parks Chairperson, Interview, Stellenbosch, 14 June 2011.

Dan, Interview, Jonkershoek Nature Reserve, 27 January 2010.

Elder Bossiedokter, Interview at his homestead, Paarl, 4 June 2010.

MTO Manager Meeting, CapeNature Conservation Offices Jonkershoek, Stellenbosch, 24 June 2010.

Ras Gad, RasTafari Priest/Bossiedokter, Paarl, 27 November 2009.

Ras Levi, Interview at medicinal garden, Franschhoek, 14 April 2010.

Ras Naphtali, Interview, Stellenbosch, 25 February 2011.

Ras Naphtali, Interview at Jonkershoek Nature Reserve, 27 January 2010.

RasTafari Bossiedokter, Interview, Idas Valley, 1 August 2010.

RasTafari Elder Man, Interview, Cloetesville, 14 May 2011.

RasTafari Elder Woman, Interview at Signing of MOU at CapeNature Conservation Offices Jonkershoek, Stellenbosch, 10 June, 2010.

Reuben, Bossiedokter, Interview, Stellenbosch, 12 January 2010.

Peanut butter salvation: the replayed assumptions of 'community' – conservation in Zambia

Elizabeth Godfrey

Department of Anthropology, University of Witwatersrand, Johannesburg, South Africa

In contemporary conservation discourse, an uncomfortable distinction exists between the foreign organisations invested in protected areas and the surrounding 'local communities'. Specifically, this dichotomy seems to reinforce a stereotypical set of imagined stakeholders, ultimately promoting Euro-American conservation experts as the instructors of uninformed residents in wild areas. This paper presents publications on two community-based conservation efforts in South Luangwa, Zambia that emphasise this division: Administrate Management Design (ADMADE) and Community Markets for Conservation (COMACO). Through community-based natural resource management (CBNRM) literature from Zambia, additional CBNRM narratives from elsewhere in Southern Africa, and limited supporting ethnographic evidence from South Luangwa, this paper demonstrates the continued Euro-American environmental hegemony of 'community-focused' conservationism. In particular, the perspectives publicised by both ADMADE and COMACO suggest that without influence from international conservation institutions, 'communities' that live amongst wildlife in Zambia will deplete their natural resources and in the process alter the African savannah as imagined and revered by the global North. This paper introduces a preliminary analysis of both ADMADE and COMACO in attempt to unveil their inherent imperial associations and highlight the local challenges to their protocols. In so doing, this critique serves as a point of departure for future comparative research on the repeat assumptions and rejections of CBNRM programmes in Southern Africa.

Good for Zambia! Good for you! Delicious peanut butter made locally in Zambia by Zambians from peanuts grown in Zambian local small-hold farms means that when you send your kids to school with an IT'S WILD 'PB&J' peanut butter sandwich for lunch, families in Luangwa Valley are finally making ends meet and able to send their own kids to school![1]

The above advertisement on IT'S WILD! peanut butter publicises how the Wildlife Conservation Society's (WCS) COMACO project has transformed the meaning of small-scale production in the Luangwa Valley, Zambia; according to COMACO literature, children have been recast as students as a result of the WCS directive to farm peanuts for profit.

COMACO is represented as a 'revolutionary'[2] initiative in the Luangwa Valley that strives to protect wildlife through an innovative market-based approach to 'communities' and conservation. The transformative power of COMACO, as evidenced by the far-reaching success of IT'S WILD! sandwich spreads, begins with the organisation of livelihoods in the Luangwa Valley. The foremost attention of COMACO is focused on poachers outside of the South Luangwa National Park (SLNP) who are encouraged to surrender their weapons and denounce hunting as a subsistence activity and commercial profession. Instead of shooting or snaring local fauna, WCS advises poachers to invest in the potential arable wealth of their surrounding land by taking up rural farming – the peanut butter business – as a means of provision. As detailed by various WCS publications, the success of the COMACO model for rural communities is profound; it has provided a space for 'reconciliation between man and nature'.[3] Specifically, it has created opportunities for 'local communities' to better their livelihoods and ensure educated futures for their children, simultaneously protecting endangered wildlife (COMACO 2010a, 2010b).

Many 'community' conversion stories in Africa celebrate the ability of international conservation organisations to reform 'local' livelihoods through economic returns from a 'sustainable' wildlife industry (tourism) and alternative natural resource use (farming instead of poaching). COMACO addresses the relationship between conservation and rural development in Zambia by improving 'the well-being of rural people while influencing land use practices' (Lewis 2010, 10). Through COMACO, WCS promotes its ability to transplant and cultivate a particular understanding of endangered wildlife and its non-consumptive value. This civilising rhetoric is not unfamiliar in Zambia; indeed, it is aligned with older conservation missions that too reproduced the colonial stage on which wise foreigners arrived to save Africa from its rampant resource depletion (and, by proxy, to save Africa from itself) (Elizabeth Garland 2008). The COMACO literature omits mention of how its poacher intervention scheme reflects past foreign-led efforts, in both South Luangwa and across other Southern African countries, to stimulate shifts in 'community' mentality.[4] By not acknowledging and absorbing the multidecade history of 'global' conservation efforts in Zambia and Southern Africa, COMACO chooses to ignore and threatens to perpetuate the studied, problematic assumptions of community-conservation initiatives.

Through a close reading of COMACO academic articles and publicity material, with minimal ethnographic attestation, this paper will investigate the prescribed 'community' in South Luangwa, Zambia and its supposed malleability as presented through the COMACO project. Furthermore, it will situate COMACO in the context of the previous WCS programme in the Luangwa Valley, the Administrate Management Design (ADMADE), and within broader critiques of community-based natural resource management (CBNRM) programmes in Zambia and Southern Africa.

The principle ambition of CBNRM programmes in Southern Africa has been to incentivise conservation with returns from sustainable land use and wildlife activities, with a particular focus on how to convince poachers to abandon their hunting livelihoods. This aspect of the COMACO project directly reflects the priorities of ADMADE both in the programmes' objectives and its 'community' audience as described and imagined. Specifically, both ADMADE and COMACO – along with many other Southern African community-based conservation programmes – failed to extrapolate on how 'the community' was interpreted within their CBNRM design. Instead, both programmes relied on homogenising descriptions and understandings of rural audiences and participants in the Luangwa Valley. The consistency of 'community' representation between many CBNRM efforts and the negative reception of these programmes in 'the community' warrant ethnographic attention. This paper, however, does not portray ethnographic findings of community-based conservation programmes in Zambia, but instead, through the literature on CBNRM in South Luangwa, Zambia, acts as an exploratory point of departure for future comparative research.

Through academic and popular publications and limited interviews conducted with South Luangwa residents in July 2012, I confront how COMACO defines and promotes its 'community' and conservation successes. Furthermore, I will consider how these representations parallel the rhetoric of other CBNRM initiatives in Southern Africa and the ways in which they epitomise the broader mission of exclusive and excluding international conservationism. In so doing, I will present the critiques of additional foreign community-conservation efforts, most notably the Communal Areas Management Programme for Indigenous Resources (CAMPFIRE), and investigate if and how conceptualisations of 'community' and 'conservation' have morphed over the decades of 'community-based' projects in Southern Africa.

Additionally, I will unveil how the continued insistence on 'the community' and its complaisance in South Luangwa is challenged and contradicted by the lived experiences of people who engage with the COMACO project. Perhaps the reactions of individuals in the Luangwa Valley to COMACO reflect their experiences with the hegemony of older 'community conservation' attempts. The political history of the Luangwa Valley reveals how and why conservation operations are met with distrust, especially when they imply the homogeneity and ignorance of rural residents – those people who are 'not yet' socially evolved to sustain and cherish their natural resources (Ferguson 1999). Finally, I will impart the modes of resistance to 'community-conservation' programmes in South Luangwa and how this discord represents larger affronts to the global powers of conservationism.

The CAMPFIRE condition

The pilot CBNRM programme in Southern Africa, CAMPFIRE, began in Zimbabwe in the late 1980s. In the 25 years since its inception, there has been extensive academic and technical critique of its theories and functions. Most notably, the way in which CAMPFIRE conceptualises and represents its 'community' audience has been repeatedly examined. The programme has been found to misunderstand what the term 'community' means in various Zimbabwean contexts (Frost and Bond 2008; Young, Makoni, and Boehmer-Christiansen 2001); its focus on one 'community' is said to occur at the expense of other local residents (Dzingirai 2003a); it operates with disregard for diverse 'community' and individual interests (Murombedzi 1999; Mutandwa and Gadzirayi 2007) and wildlife values (Virtanen 2003); and the programme fails to engage 'communities' effectively (Alexander and McGregor 2000; Dzingirai 2003b), especially the non-elite residents (Wolmer, Chaumba, and Scoones 2004). The homogenising outlook applied by CAMPFIRE has also been noted in several other CBNRM programmes in Southern Africa, which too tend to oversimplify and neglect the 'real institutional matrix' that comprise 'communities' (Leach, Mearns, and Scoones 1999, 240), resulting in 'significant confusion' over how the programmes are supposed to function and for whom they are meant (Musumali et al. 2007, 306).

Furthermore, the ethos and operations of CBNRM programmes and conservation initiatives in Africa more generally have been scrutinised for their 'blatant recycling of retrograde colonial fantasies' that celebrate and elevate the foreign wildlife experts and developers above local participants and residents (Elizabeth Garland 2008, 58). In Zimbabwe, Zambia, and elsewhere in Southern Africa,[5] CBNRM programmes are meant to serve as 'antidote[s] to rural poverty', but have been shown to ineffectively balance the scientifically 'confirmed' ecological concerns[6] with socioeconomic circumstances (Logan and Moseley 2002, 1). Rather, the attention of the programmes has concentrated on the technical issues of implementation at the expense of the relevant historico-political contexts (Alexander and McGregor 2000; Nelson and Agrawal 2008). The widespread insistence on 'traditional' leaders as liaisons and representatives for the 'community' has furthered the gap between CBNRM promoters and the ordinary residents of the areas in which these programmes persist (Kumar 2005; Marks 2001). This increasing divide has resulted in deep distrust of CBNRM methods and direct (sometimes violent) challenges to its framework and everyday operations (Gibson and Marks 1995).

Designs for 'community' management in Zambia

Residents in the Luangwa Valley have a potent reaction to the term 'conservation', largely due to *Save the Rhino*, an Non Governmental Organisation formed in the 1980s to secure the nearly extinct black rhino population in the SLNP. The organisation was formed by a group of white safari guides and tour operators in South Luangwa, who travelled to Europe, the United States, and Canada to give talks at prestigious universities, broadcasting well their plight to secure the remaining Zambian rhinos (S. Marks, pers. comm.). They gained a substantial support base; but, in the end, the organisation was not able to save the rhinos and the species vanished in the late 1980s (Astle 1999). As a result, the association between a conservation organisation and the impending disappearance of its subject species was – and still is – highly suspected by Zambian public.

In response to the loss of rhinos, and its implied prognosis for the extended demise of Zambian wildlife, other international organisations became involved to help salvage what was left of Zambia's wilderness. Two community-based natural resource management (CBNRM) projects began in the 1980s: the Administrative Management Design (ADMADE) and the Luangwa Integrated Resource Development Programme (LIRDP). They were funded, respectively, by United States Agency for International Development (USAID) in partnership with the WCS and the Norwegian Agency for Development Cooperation (Norad). Individually, these programmes tried to incentivise conservation by availing a portion of safari hunting profit, by various means, to the communities around South Luangwa. Both programmes persisted into the 1990s, with the ethos of LIRDP eventually incorporated into the formation of Zambia Wildlife Authority (ZAWA) in 1998 (Wainwright and Wehrmeyer 1998). Although these community-conservation approaches appeared in reaction to the loss of rhinos, they were met by Zambian dubiousness and distrust of the agents and mechanisms behind conservation agendas.

> When LIRDP came in the 1990s, they talked to villages, they used drama and other means, with the message that conservation can bring money; they had printouts and drawings and they told us that we own the animals. Those were better days for conservation. LIRDP gave cash to anyone with a national registration card; I got 14,000 ZMK and I bought 2 KGs of sugar. This was a time of heavy poaching and people had an evil feeling about conservation. The message of LIRDP was that these are your animals, you should get the benefits. Giving cash was the fastest way to get people's attention.[7]

Dennis Phiri, ZAWA Assistant Ecologist for SLNP, described for me some of the first public uses of the word 'conservation' by foreigners in the Luangwa Valley, specifically LIRDP. The financial support for LIRDP was eventually blended with the wildlife management unit when it transitioned from National Parks and Wildlife Service (NPWS) to the ZAWA in the late 1990s. LIRDP was subsequently criticised for presenting residents around the South Luangwa National Park with cash handouts; but, as Dennis explains, the approach was a direct method to instil in people the correlation between conservation of wildlife and monetary rewards.

Beyond the methodology, LIRDP delivered a powerful message: these are your animals and you must benefit from their conservation. Benefits were defined by LIRDP as immediate remuneration and the concept made an impact; almost 20 years later, Dennis can still remember exactly how much money he received and what he purchased. Effective conservation can sweeten your existence; this was the sentiment promised by community-based approaches to conservation. But the strategy did not reflect the everyday outcomes of wildlife conservation; people do not see these kinds of material returns on a regular basis or in such directness. Instead, the 'benefits' come in the form of limited job creation through a market for healthy wildlife that is appreciated by international tourists. By not killing the animals, by ensuring that their habitats and corridors remain intact, the residents of the Luangwa Valley are said to contribute to the long-term survival of their species and landscapes. The longer these populations remain, especially those that attract the most attention and the most income, the more revenue will flow through South Luangwa. With more money available in the area, the chance that it will manifest in people's everyday lives is greater.

Administrating 'community'

The ADMADE, originally known as the Lupande Development Workshop, began in 1987 and ran concurrently with LIRDP. This CBNRM project was funded by the USAID, in partnership with WCS. The formation of ADMADE was in response to the Lupande Development Workshop in 1983, a workshop that was organised by the Zambia National Park and Wildlife Services.[8] ADMADE recognised 'the need for local communities to have stewardship over and receive economic benefits from sustainable use of wildlife resources' (Clarke 2000) – a new participatory approach to conservation that reflected the international mandate to link conservation and development (IUCN 1980) and paralleled other pilot approaches in Southern Africa, most notably CAMPFIRE (Chambers 1983;

Young, Makoni, and Boehmer-Christiansen 2001). The focus of ADMADE was initially the distribution of revenue from the safari hunting industry, specifically the allocation of a percentage of this revenue to 'the community' through a 'community account' (known as the Wildlife Conservation Revolving Fund). In addition to revenue through the Revolving Fund, ADMADE employed village scouts for anti-poaching patrols and opened a training centre known as the Nyamaluma Institute. Nyamaluma was primarily used as a lab to generate geographical information system (GIS) data and maps, with the goal 'to help improve the capacity of rural communities to become more knowledgeable and effective in managing their wildlife resources' (Lewis 1995). ADMADE also implemented what was known as 'Green Bullet certification' for hunting safari operators, an agreement meant to symbolise 'partnership between the safari operator and the local community according to ADMADE guidelines' (Lyons 2000, 40).

The technical advisor for WCS, Dale Lewis, explained the programme as one that stressed 'the benefits to and responsibilities of local residents who are prepared to share their lands with wildlife' while respecting the 'ethnogeography and political system' of its participants (Lewis 1989). Respect for the political system in the context of ADMADE meant working through a certain understanding of African land use practices: 'To rural Africans...communal customs governing land use... are an accepted fact' (Lewis 1995, 861). A clear distinction was made between western 'free market concepts' and the practices of rural Africans; however, due to the influence of an 'overly centralised government', the ADMADE literature acknowledged that new political lines were re-shaping rural African structures from 'old to new...from African to more western values' (861). Through the depiction of a politically evolving rural Africa, ADMADE aligned itself with 'traditional' principles, the pre-western version of African governance, and claimed to operate with 'an appreciation for African customs' that other foreign-based projects ignored (Lewis 1995). In practise, their method was to work through chiefdoms to distribute funds and use GIS applications to identify 'resource management needs and appropriate responses by the community to such needs' – a 'bottom-up' approach to community research (863). The GIS data was then presented to 'community leaders' to ensure that the information reached the larger 'community' and it was used to stimulate discussions about future management needs (Lewis 1995).

Throughout USAID funding to ADMADE, mismanagement of money by various chiefs resulted in hunting returns never reaching the non-ordinary village participants. Additionally, the GIS information was not communicated to 'the community' at large (Clarke 2000; Gibson and Marks 1995; Nelson and Agrawal 2008). Furthermore, ADMADE was designed to work within 'traditional' authority structures, resulting in the alienation of the broader and diverse 'community' (Marks 2001).[9] By attempting to work within these existing authority structures, ADMADE relied on a specific, older concept of rural Africa as disengaged from the global free market and involved in 'traditional' communal land use customs (Lewis 1995). After ten years and a total of US$ 4.8 million worth of investment (Clarke 2000), USAID stopped its funding to ADMADE in 1999.

There have been several critiques of the way in which ADMADE began and ended in the Luangwa Valley (cf. Marks 2001; Lubilo and Child 2010; Wainwright and Wehrmeyer 1998), but what has not been thoroughly analysed is how the programme conceptualised its audience. The broad idea of ADMADE was to redistribute revenue from the safari hunting industry to benefit people in the Lupande Game Management Area (GMA) more directly. While the programme claimed 'its very foundation' as 'the local community' (Lewis 1989, 3), it did not define exactly what was meant by this phrase, that is, who comprised 'the community', or how contemporary transitions 'from...African to western' were negotiated by ADMADE in the context of the Luangwa Valley. The assumption that hunting revenue would be evenly distributed through the chiefs proved unsuitable and how this democratic element fit within the communal design of ADMADE was never proposed. Not until 1998, through the ZAWA Wildlife Act, was this structure rearranged, establishing an elected Community Resource Board to manage tourism revenue with the chief acting as merely an honorary patron.[10]

Additionally, ADMADE was found to support many 'free riders' who would 'continue to hunt while receiving benefits' from ADMADE officials (Gibson and Marks 1995, 942). These free riders were able to both behave publicly according to ADMADE expectations and also covertly persist in direct wildlife use. The ADMADE literature did not account for the historical and symbolic role of hunters in South Luangwa, especially as the tsetse flies in this region precluded the keeping of livestock. Towards the end of ADMADE, the known illegal hunting off-takes were equivalent to, if not greater than, those prior to ADMADE, with the increasing usage of snare wire helping to further conceal the vocation (Gibson and Marks 1995).

As one of the first CBNRM attempts to follow the well-known and much criticised CAMPFIRE programme, oversights in the ADMADE framework were inevitable. Yet despite stating that ADMADE was run with an appreciation of Zambia's ethnogeography, exactly how this appreciation was learnt and what consciousness of 'ethnogeography' meant amidst changing political structures was not apparent. The act of monetary allocation through the chiefs and the idea that funds should be distributed to 'the community' in a manner that reflected 'African customs' were not just reflective of CBNRM in its infancy. Rather, this simplistic system was designed to engage a 'mythic community' – an imagined structure that was never considered for the diversity within and between its various members (Agrawal and Gibson 1999). More specifically, the system neglected 'the differential access of actors' within the Luangwa Valley 'to various channels of influence, and the possibility of 'layered alliances' spanning multiple levels of politics' (Agrawal and Gibson 1999, 640). Through the language of ADMADE, this mythic community was described not only as uniform, but also as a pre-modern, pre-western structured society – rhetoric that is in line with other community-based projects in Southern Africa and overtly reminiscent of colonial depictions (Alexander and McGregor 2000; Garland 2008; Young, Makoni, and Boehmer-Christiansen 2001).

The importance of 'the community' to conservation developed as a particular discourse for CBNRM and as a means to recruit local resource users into Euro-American anxieties about biodiversity conservation and wildlife protection. As a whole, the initial language for CBNRM activities relied upon a particular kind of representation, one that was faulted repeatedly in the CAMPFIRE context (cf. Alexander and McGregor 2000; Dzingirai 2003a; Frost and Bond 2008; Wolmer, Chaumba, and Scoones 2004) and eventually ruptured the ADMADE design. The insistence on a dichotomous division – African versus western, old versus new – reinforced stereotypical, colonial characterisations, and instead of working with and through the acknowledged changing political structures, ADMADE catered to an anachronistic understanding of 'traditional' African governance.

'Global' conservationism

Today, organisations like WCS enter into community-conservation 'partnerships' as the international enforcers of specific kinds of acceptable 'community' wildlife usage defined by the global conservation decision makers. WCS and other leading environmental institutions (e.g. International Union for Conservation of Nature (IUCN), Convention on International Trade in Endangered Species (CITES), Conservation International (CI), World Wide Fund for Nature (WWF)) are meant to provide local governments and wildlife management bodies with recommendations on how to establish 'conservation priorities' (Ceballos et al. 2005, 603). In other words, they initiate the sociobiological groundwork from which larger policies and schemes for environmental protection are to be constructed. The local governments, with support from major international funding bodies like USAID and the World Bank,[11] serve as the primary national supporters to ensure that such policies are carried out (Frank, Hironaka, and Schofer 2000). In so doing, the nation state cooperates within a particular discourse of environmentalism, one that asserts the importance of the 'global' benefit of effective local conservation (Frank, Hironaka, and Schofer 2000; Young, Makoni, and Boehmer-Christiansen 2001). This discourse of environmentalism, which now includes the essential 'community' component, is also dictated by the impending urgency of saving certain species and spaces as categorised on CITES red lists (Rodrigues 2006). The level of anxiety about protecting endangered fauna correlates to their geographical locations and direct action is prioritised for countries that are 'identified as being most at risk, having both exceptionally high richness and endemism and exceptionally rapid rates of anthropogenic change' (Ceballos et al. 2005, 603). In countries like these 'an unprecedented international effort will be needed – one requiring the development of both new attitudes and institutions', (606). The global must supersede local needs to enable international conservation institutions to 'stimulate' local communities and wildlife authorities 'by providing a big-picture perspective on the current and projected status of biodiversity on the planet' (Ferrier et al. 2004, 1101).

This 'global' concern over the protection of biodiversity and the 'attitudes' required to effectively conserve vulnerable fauna are communicated daily through organisations like WCS that are commissioned to protect Nature by convincing local 'communities' of their global worth (Tsing 2003, 164). WCS represents this global agenda directly through their projects and publications, which 'make claims for nature and for the globe' through universalising terms that purport to benefit 'communities' (Tsing 2005, 112). The rhetoric of community-conservation projects makes it extremely difficult to discern when a 'community' is considered responsible enough to direct its own decisions about wildlife and when a country's conservation policies qualify the nation as environmentally modern. Revised and effectively implemented community-conservation policies represent more than just national cooperation and 'respect' for Nature; they also demonstrate the social evolution of the country and its 'communities' towards the new modern 'attitudes' of sustainable living (Spaargaren, Mol, and Buttel 2000).

CBNRM programmes are designed to 'help' post-colonial 'communities' and countries to manage their wild areas for local and 'global' good. In the process, however, these programmes inherently dispossess local citizens and their countries of the sovereign right to categorise wildlife protection according to domestic values. Instead, decisions about the use of wildlife and how it should be appreciated are pre-determined by the major Euro-American conservation organisations, which rely on ecological science to confirm the fragility of specific wild populations. These decisions are then delivered to the post-colonies in the guise of programmes that promise to serve the interests of people, but constantly emphasise Northern philosophies on wildlife worth and management. The resulting suspicions of the global community-conservation agendas appear in unexpected, peculiar forms and challenge the disguised agents of the environmental mission to Africa (Chakrabarty 2009; Garland 2008). In South Luangwa, as a means to engage the 'evil feelings about conservation' that Dennis described, Zambian public make use of 'occult cosmologies' to resist opaque CBNRM and other conservation programmes (West and Sanders 2003, 6; Comaroff and Comaroff 2003). Initially, these cosmologies were directed at *Save the Rhino* through rampant suspicions and accusations of possible illegal and illicit components of their conservationism. However, the anxieties related to wildlife conservation have persisted beyond *Save the Rhino* and have also been directed at CBNRM programmes in Zambia (Gibson and Marks 1995). The stories of conservation disingenuousness point and prod at the unknown, forceful mechanisms behind an environmentalism that supposedly benefits the primary wildlife users. They signify that, as a result of globalising conservation efforts, Zambia and Zambians are deprived of something that is difficult to name. Instead, their disquiet is communicated through accessible metaphor; *Save the Rhino* took the most valuable species away from Zambia, and neither ADMADE, LIRDP, nor any other conservation initiative has been able to rectify the collective sense of injustice.

The wild assets of South Luangwa

The SLNP is the most economically productive and frequented national park in Zambia, with 25,393 foreign tourists and 6994 resident and citizen tourists in 2011 – a 16% increase in visitor numbers from 2007.[12] The GMA to the east of the park, the Lupande GMA, spanning an area of 4800 km^2, recently received an estimate of close to US\$500,000 per year in revenue from hunting safaris, in addition to the \$1,000,000 revenue from park fees.[13] The park is the second largest in Zambia, approximately 9050 km^2 in size, and 'one of the top 10 best-known and most popular tourist destinations in Africa owing to its abundant wildlife, scenic beauty and wilderness state' (Zambia Wildlife Authority 2011). The park receives foreign tourists throughout both the wet and dry seasons, most of whom partake in photographic safaris with a small percentage involved in consumptive (hunting) tourism. There are 13 lodges, 23 bush and mobile camps, and 3 hunting lodges in the park and Lupande GMA, all with their own facilities and staff. The result of tourism expansion in South Luangwa and the subsequent increase in employment opportunities is said to have tripled the Lupande population in the last twenty years, with an estimated 60,000 people now residing on the park's borders.[14] Of this population, roughly 1000 people are employed directly through the lodges and camps (1.67% of Lupande residents), with others hired on contracts or ad hoc bases (Figure 1).[15]

Businesses in the Lupande GMA cater to the industry; shops are constructed to serve various lodge departments, formal and informal trainings for young men who want to be safari guides have been established, and various school programmes and NGO-organised events reinforce the monetary and inherent value of wildlife. In the 1980s, the last of the black rhino population in South Luangwa was commercially poached to extinction, leading to an intensive anti-poaching effort to stabilise the declining elephant numbers.[16] While the numbers of elephants are said to have

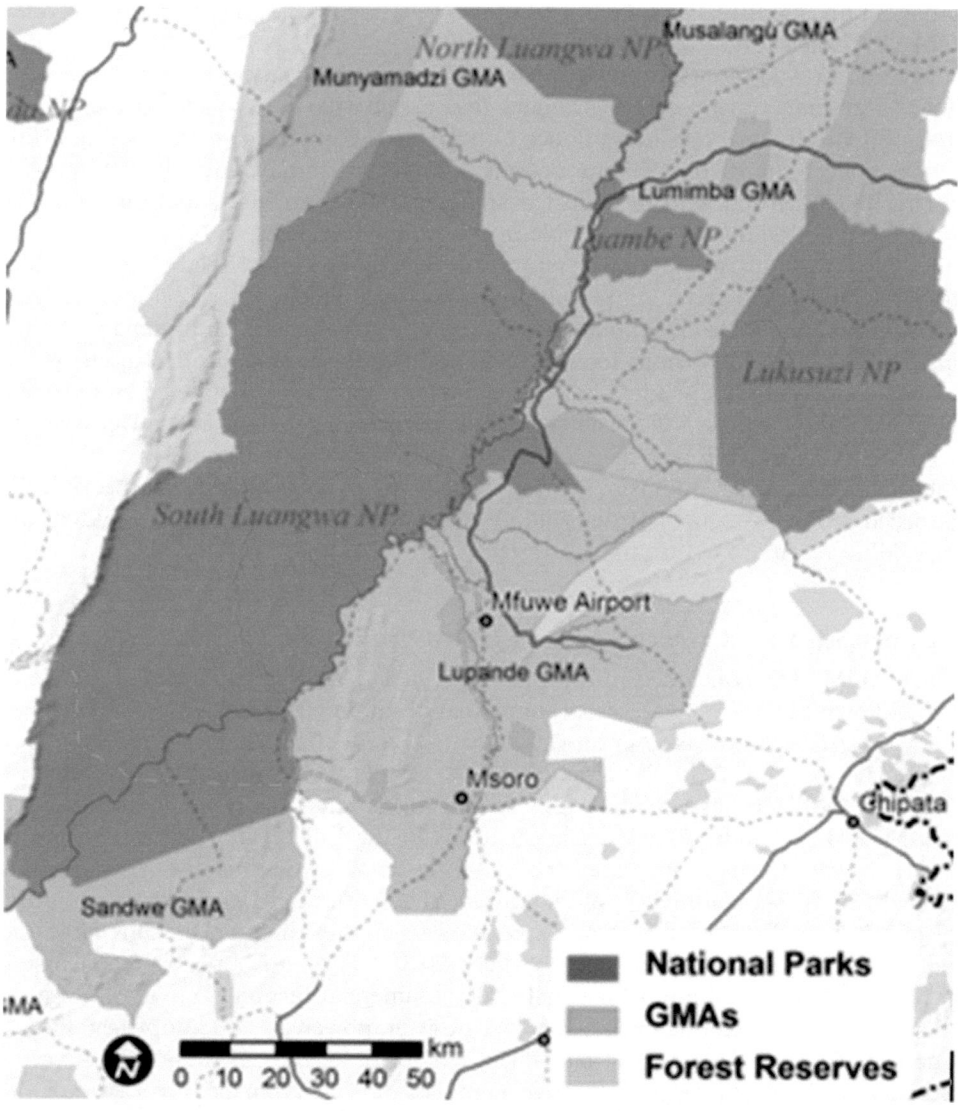

Figure 1. Map of South Luangwa National Park and the Lupande GMA. Map courtesy of Dr. Fred Watson.

decreased drastically, the surge of anti-poaching effort in the 1980s and 1990s was successful insofar as it secured the remaining population. Today, the individuals employed in anti-poaching belong to two groups: the ZAWA, which is the management organisation for most of the protected areas and sanctuaries in Zambia, and village scouts, who are employed through their respective Community Resources Boards to monitor illegal hunting activities. Additionally, the South Luangwa Conservation Society, an NGO devoted to anti-poaching and law enforcement, supports both ZAWA and the village scouts logistically and financially. These various scout assemblies cooperate under formalised agreements and collectively symbolise the fragility and wealth of the Luangwa wildlife. Regardless of how one measures anti-poaching success, the structures that have emerged since the extinction of rhino represent an extensive symbolic boundary between the illegal hunters in the Lupande GMA and their target species.

In addition to scouts, there are also NGOs (e.g. Chipembele Wildlife Education Trust, Zambian Carnivore Programme) that assist with anti-poaching; there is an assembly of Honorary Wildlife Police Officers that conduct patrols; walking safari guides collect snares on their excursions; an 'ex-pat' initiated jewellery trend has emerged to turn the wire of collected snares into adornments ('snare ware'). The significance of wildlife is unavoidable in the Lupande GMA, and one is constantly inundated with images of open safari vehicles (packed with elite, mostly white tourists) and dark green scout uniforms, both of which reiterate the wealth of the non-human. Since the extinction of the rhino, residents in the Luangwa Valley have been named by various international community-conservation projects as the caretakers of Zambian wildlife and the ultimate deciders of the fate of biodiversity. However, the notions of ownership and participation that are advertised by community-conservation projects are managed within a particular paradigm of international conservationism, within which 'the community' is theoretically afforded monetary benefits from wildlife tourism but not the direct, individual use of wildlife resources (Murombedzi 1999).

New channels for old assumptions: the COMACO initiative

Once ADMADE concluded in the Luangwa Valley, WCS began a new initiative in 2002 known as COMACO, a programme that is still in effect today. The objective of the COMACO project is to improve biodiversity and 'effect change at an ecosystem scale' (Lewis et al. 2011) by amending how people utilise resources. The premise behind COMACO, as stated by the WCS Programme Officer in Zambia, is that 'COMACO will stabilise wildlife populations in the Luangwa Valley' (Lewis et al. 2011, 13959) through increased potential in small-scale farming. To test this hypothesis, WCS identified the main household activities for 'Luangwa communities' as poaching and subsistence farming, both of which they qualified as 'unreliable and unhealthy' (Wildlife Conservation Society 2010). Through a poacher conversion programme and facilitation for small-scale commercial farming activities, however, 'communities' in the Luangwa Valley are now 'empowered ... to adopt new trades' and simultaneously the project has 'taken tens of thousands of snares and firearms out of the bush, saving equally large numbers of wild animals'.[17] As part of the COMACO programme, WCS invites donors to 'Adopt a Poacher!' for US$350, which will 'transform a poacher' through a six-week course on 'alternative trade skills' and 'help save an estimated 23 animals per year, the average number of animals a poacher hunts illegally every year in Zambia', (Wildlife Conservation Society 2010). According to a 2010 analysis of the COMACO programme, their hypothesis is proving true: 'Biodiversity conservation was the major underlying motivation for COMACO, and most species show stabilisation of the populations relative to the declines noted in the 1990s' (Lewis et al. 2011).

The story of a poacher relinquishing his weapons[18] is one that appeals to a particular kind of donor audience and celebrates the capacity of foreign organisations to permanently influence rural livelihoods and save the wild areas of Africa. COMACO offers a slightly different approach to ADMADE; instead of relying on chiefs as the intermediaries, COMACO engages 'the community' directly, creating a safe space for poachers to surrender their former livelihoods and assume new, more lucrative professions. Yet while the lines of communication have been re-evaluated post-ADMADE, the same implicit assumptions can still be read through the COMACO literature, namely that a foreign organisation has the dexterity to 'transform' the way in which people value wildlife. Specifically, 'the poacher' is presented as a fixed figure who is ignorant of farming livelihoods and in need of edification. 'WCS studies revealed that without aid, most farmers in the Luangwa Valley experience three to five months of chronic food insecurity' (Wildlife Conservation Society 2010); COMACO therefore teaches how to both initiate farming practices and also sustain them. The emphasis of COMACO is on foreign aid to 'a community' that needs assistance, and the aid is administered in a way that places people at the forefront of a conservation programme that is designed to protect wildlife.

Other representations of the Luangwa Valley are quite different from the picture portrayed by WCS. Tourism has grown substantially since the mid-2000s, and no lodge in the Park or the Lupande GMA has ever closed due to lack of business.[19] ZAWA currently has more applications for new concessions than they can manage (D. Phiri, pers. comm.). Beyond tourism, ZAWA, the South Luangwa Area Management Unit (SLAMU), government structures, local schools, shops, the international airport, and other opportunities are also expanding in the Lupande GMA. The recent decline in tsetse flies – a manifest benefit from reduced wild populations – has resulted in an increase of livestock and cotton production for some farmers. There are not, however, endless job possibilities for Lupande residents; poverty is still prevalent and there are never enough jobs to satisfy demands for employment. But the Lupande GMA is not devoid of wealth, farming opportunities are broadening, and there are multiple government and other Zambian-led programmes to address issues of food security and poaching. The characterisation of a famished, forgotten 'community' as portrayed through COMACO literature justifies a foreign-led mission to save rural Zambia, mirroring the image of pre-modern, undemocratic Africa that was promoted by ADMADE to rationalise its assistance.

The result of this interpretation for ADMADE was local collusion to re-appropriate international funds. For COMACO, 'reformed poachers' are said to fashion fake muzzle loading guns and construct snares through stolen COMACO wire, which are presented to COMACO officials in exchange for farming tools and seeds.[20] These materials are then used to further the subsistence farms that many of these 'converted poachers' have long managed (ZAWA scouts, pers. comm.). Furthermore, the rates of poaching and snaring in the Luangwa Valley continue to increase (Becker et al. 2013). ZAWA officials in South Luangwa describe COMACO as an extension of ADMADE and struggle to understand how the advertised benefits of the programme, the sale of products to high value markets, render the COMACO participants better off than other Luangwa farmers (ZAWA scouts, pers. comm.).

What emerges through these testimonies about COMACO is that its benefits to 'the community' and relationship to conservation are not transparent nor are they readily understood. Additionally, the COMACO protocol has been covertly super-seded by another local market: one for artificial arms that mocks the assumption of 'community' transformation. This manipulation reveals the defective philosophy with which many community-conservation organisations operate. 'The poachers' outside the SLNP discredit COMACO through their circumvention of its protocol and belie the representations of starving, compliant rural people. The wildlife management body is sceptical of its usefulness and critical of COMACO methods (ZAWA scouts, pers. comm.). The more that organisations like WCS emphasise the harm of poaching for both local residents and the wildlife-loving world at large, the more this form of off-take seems to increase.[21] The COMACO programme relies on a particular understanding of wildlife and its value, one that does not account for the political, historical, and lived diversity amongst residents of wild areas. This inherent disconnectedness results in repeated rejection by local individuals who witness the latest recycled community-conservation attempts in action.

Conclusion: future implications

The messages of CBNRM programmes specify that residents along protected areas ultimately have control over their surroundings; ultimately they are the owners and managers of their wildlife. They do not, however, extrapolate the limits to the definition of 'ownership' for both the people who live with wildlife and the countries that house endangered species. In the most conventional understanding of the noun, an owner has sovereign choice over that which she/he possesses. But for CBNRM, the notion of 'ownership' is managed within a particular framework, delineated by the canon of international conservation, which a priori dispossess any one person, community, or country 'ownership' of 'global priority' spaces and species.[22] This contradictory presentation of CBNRM is easily unveiled and perhaps explains why community-conservation programmes are met with strong scepticism and frequent dismissal. Yet, through the self-narratives and critiques of CBNRM, the disconnect-edness between these programmes and the larger implications of their objectives becomes apparent. Most notably, effective cooperation with CBNRM programmes necessitates a shift in the historically and culturally constructed mindsets of the many individuals who comprise rural populations. The programmes assume that rural Africans can simply be 'transformed' to think like Euro-Americans as long as transformations come with sufficient incentives. As exemplified through both the critiques of CAMPFIRE and the focused attention on ADMADE and COMACO, blind acquiescence never happens. Instead, residents of wild areas devise creative, impassioned schemes to re-assert the ownership and rights that CBNRM pro-grammes try to covertly disaffirm (Dzingirai 2003b).

Through the literature of ADMADE and COMACO, I have attempted to unravel how the rhetoric of these projects assumed a particular kind of audience, with one singular, amassed identity.[23] The language of those assumptions was used to portray the Luangwa Valley, Zambia as a place in need of foreign attention to manage the human population and the effects of this population on the surrounding wildlife. Through the focus on 'traditional rural Africa', ADMADE did not unfold in the prescribed fashion; on the contrary, their attention to 'African customs' resulted in the alienation of individuals outside the political arena. Similarly, the COMACO insistence on poacher transformation assumes a kind of pliancy that has been shown untenable. While the two programmes took slightly different approaches, the literature of both ADMADE and COMACO disregarded the contexts of conservation in South Luangwa and potential existing appreciations for wildlife. Both insisted on the vulnerability of African people and African non-human species, a condition that required foreign intervention.[24] In the Lupande GMA, both of these programmes have been exploited in unprecedented ways, which highlights the outdated, unproductiveness of imperial perspectives. Instead of engaging the studied setbacks of other CBNRM programmes in Southern Africa, ADMADE and COMACO publications addressed almost exclusively the methods and evidence of their own successes, choosing to disregard the contexts and complications that shaped their CBNRM applications. Had they done so, they might have complemented other community-conservation critiques and added to a body of literature that highlights where and how such programmes fall short of expectations. The absence of critical self-reflection and dialogue with other CBNRM obstacles, however, results in repeated misassumptions, lingering colonial attempts to manage Africans and their land, and evokes further rumours that foreign-led conservation schemes serve only the interests and perspectives of those who initiate them.[25]

While there have been several critiques of various CBNRM programmes, with consistent attention to their inadequate knowledge and conception of imagined 'community' audiences, there is an overall lack of research into the pervasive overlap of these deficiencies in Southern Africa. Moreover, the local rejections and manipulations of a number of CBNRM protocols have yet to be thoroughly scrutinised in a comparative ethnographic study of how community-conservation programmes elicit similar strategic responses from their targeted audiences. There have been comparative studies of how different 'communities' react to the same CBNRM programmes (Alexander and McGregor 2000; Dzingirai 2003a, 2003b) and how distinct CBNRM efforts have impacted residents in a defined geographical area (Nelson and Agrawal 2008); but the regional and perhaps continental strategies to combat the ethos of 'community'-focused conservation could be extrapolated in a comprehensive ethnography. Through such a study, the ways in which individuals and groups of people make sense of CBNRM missions, and how they assert their non-conservationist priorities, would be presented alongside one another in a format that might be more difficult for international conservation organisations to ignore. Furthermore, such research might encourage these worldwide organisations to rethink and reshape the inherent implications and tactics of their CBNRM designs. Incorrect assumptions of 'community' participants as units of people without personal and powerful conservation experiences would be recognised as a strong component to CBNRM failure, and this admission might allow for reflexive awareness of the (colonial-style) damage that can be caused by international conservationism.

IT'S WILD! peanut butter may be delicious and COMACO may have introduced this item to a broader Zambian audience; but farming ground nuts and the sale of homemade peanut butter are well-developed practices in the Luangwa Valley. People who choose to hunt illegally are not starving individuals who are uninvolved with small-scale farming endeavours (Gibson and Marks 1995). In the post-colonial world of African conservation, rampant CBNRM misrepresentations and problematic perspectives on 'the community' fail to portray the complexity and ownership that may exist within and between bio-diverse and 'important' wild places.[26] In so doing, an outdated image of vulnerable rural Africa is perpetuated abroad to the people and organisations that fund community-conservation programmes, despite the confrontation and rejection that can develop on the ground. Why does each new 'community' conservation model seem to reproduce the studied ignorance of the previous one? The blatant reuse of ADMADE rhetoric by COMACO, and the ways in which both programmes have been rejected in the Luangwa Valley, are perhaps the first steps in tracing the Southern African network of outdated and ill-fitting CBRNM assumptions and effects. Specifically, attention should be placed on how imposed conservation philosophies are re-negotiated and controlled by the very individuals for whom such programmes are designed to pacify and reform.

Acknowledgements

I would like to thank my research informants in South Luangwa, especially Dennis, for their thoughtful, patient reflections. I would also like to extend a special thank you to the organisers of the Old Land New Practices conference for their support of this paper and for challenging me to develop its argument and analysis.

Notes

1. http://www.itswild.org/peanut-butter.
2. National Geographic Video about COMACO, www.itswild.org.
3. http://www.itswild.org/maps-cant-talk-and-are-we-listening-anyway.
4. Agrawal (1995, 2002) and Scott (1998) confront representations of the 'community' as an abstracted, yielding unit of people, with specific reference to the inadequacies of development philosophies and agendas.
5. See Twyman (2000) and Mbaiwa and Stronza 2011 for studies on CBNRM in Botswana and Long (2004) and Jones and Weaver (2009) for perspectives from Namibia.
6. Science-based management, with commitment to stand-alone facts is the governing voice behind implementing many conservation policies, see Berkes (2004) and Gunderson and Holling (2002).
7. Interview with Dennis Phiri, Assistant Ecologist for ZAWA in South Luangwa, at the Chinzombo Research Station, July 2012.
8. See Lyons (2000) for thorough review of ADMADE and the lessons learned from its published successes and failures.
9. Similar research was done on CAMPFIRE, specifically the programme's exclusive focus on 'producer communities' at the expense of other communities and individuals that were equally affected by wildlife, see Dzingirai (2003a).
10. Zambia Wildlife Authority, The Zambia Wildlife Act: 1998.
11. The World Bank hosts the Global Environmental Facility (GEF), which contributed US$4.5 million to CAMPFIRE. See Young, Makoni, and Boehmer-Christiansen 2001 for discussion on how such 'global' and powerful funding bodies work to claim a 'monopoly on the wisdom to manage nature in the South' (302).
12. Figures courtesy of the Zambia Wildlife Authority, Chinzombo Research Station.
13. South Luangwa Area Management Unit Reports, 2007 and 2008.
14. Figures courtesy of the Central Statistics Office, Zambia.
15. Figures courtesy of the Luangwa Safari Association.
16. Jachmann and Billiouw 1997.

17. http://www.wcs.org/conservation-challenges/local-livelihoods/farming-communities/wcs-and-comaco.aspx.
18. Mackenzie (1988) and Steinhart (2006) discuss the hunter and the poacher as figures in colonial and post-colonial Africa, specifically the forced adjustment of Africans from the former to the latter.
19. Interview with Zambia Wildlife Authority officials, Chinzombo Research Station.
20. Gibson and Marks 1995 also discuss similar 'free rider' manipulations in the context of ADMADE.
21. Becker et al. (2013) discuss the increased level of poaching in the Luangwa Valley, specifically the rampant use of snare wire.
22. Wildlife Conservation Society 'Saving Wildlife' mission statement: http://www.wcs.org/saving-wildlife.aspx.
23. See Scott (1998) for a persuasive attack on development theory and the violating assumptions of similitude.
24. Euro-American conservation interventions often arrive with particular neo-liberal agendas and paternalistic instructions on how 'the locals' can become globally-informed citizens, see Ferguson (2006) and Aronowitz (1988).
25. And, moreover, to further align these activities with the colonial rhetoric from which they claim practical distance. Stafford (1989), Irwin (1995), and Bajaj (1988) explore these contradictions in the context of foreign scientific paternalism – a useful perspective as most (if not all) foreign conservation efforts arrive with scientific warrants.
26. Wildlife Conservation Society 'Saving Wildlife' mission statement: http://www.wcs.org/saving-wildlife.aspx.

Notes on contributor

Elizabeth Godfrey is a Master's student at the University of Witwatersrand in the department of Anthropology. Her dissertation explores questions of citizen and state sovereignty in the context of international conservationism in Zambia. She can be contacted at eagodfrey@gmail.com

References

Agrawal, Arun. 1995. "Dismantling the Divide Between Indigenous and Scientific Knowledge." *Development and Change* 26 (3): 413–439. doi:10.1111/j.1467-7660.1995.tb00560.x.

Agrawal, Arun. 2002. "Indigenous Knowledge and the Politics of Classification." *International Social Science Journal* 54 (173): 287–297. doi:10.1111/1468-2451.00382.

Agrawal, Arun, and Clark C. Gibson. 1999. "Enchantment and Disenchantment: The Role of Community in Natural Resource Conservation." *World Development* 27 (4): 629–649. doi:10.1016/S0305-750X(98)00161-2.

Alexander, Jocelyn, and JoAnn McGregor. 2000. "Wildlife and Politics: CAMPFIRE in Zimbabwe." *Development and Change* 31 (3): 605–627. doi:10.1111/1467-7660.00169.

Aronowitz, Stanley. 1988. *Science as Power: Discourse and Ideology in Modern Society*. Hampshire: Macmillan Press.

Astle, William L. 1999. *A History of Wildlife Conservation and Management in Mid-Luangwa Valley, Zambia*. Bristol: British Empire and Commonwealth Museum.

Bajaj, Jatinder K. 1988. "Francis Bacon, the First Philosopher of Modern Science: A Non-Western View." In *Science, Hegemony and Violence*, edited by Ashis Nandy, 24–67. Tokyo: The United Nations University.

Becker, Matthew, Rachel McRobb, Fred Watson, Egil Droge, Benson Kanyembo, James Murdoch, and Catherine Kakumbi. 2013. "Evaluating Wire-Snare Poaching Trends and the Impacts of by-Catch on Elephants and Large Carnivores." *Biological Conservation* 158: 26–36. doi:10.1016/j.biocon.2012.08.017.

Berkes, Fikret. 2004. "Rethinking Community-Based Conservation." *Conservation Biology* 18 (3): 621–630. doi:10.1111/j.1523-1739.2004.00077.x.

Ceballos, Gerardo, Paul R. Ehrlich, Jorge Soberón, Irma Salazar, and John P. Fay. 2005. "Global Conservation: What Must We Manage?" *Science* 309 (5734): 603–607. doi:10.1126/science.1114015.

Chakrabarty, Dipesh. 2009. "The Climate of History: Four Theses." *Critical Inquiry* 35 (2): 197–222. doi:10.1086/596640.

Chambers, Robert. 1983. *Rural Development: Putting the Last First*. Essex: Pearson Education.

Clarke, J. E. 2000. *Evaluation of Wildlife Conservation Society's Administrative Management Design Project (WCS/ADMADE)*. Zambia: United States Agency for International Development.

COMACO. 2010a. "Peanut Butter|Community Markets for Conservation." Accessed July 15, 2012. http://www.itswild.org/peanut-butter

COMACO. 2010b. "COMACO-Sustainable Food Security, Wildlife Conservation, & Ecosystem Preservation in Zambia Africa." Accessed July 15, 2012. http://www.itswild.org

Comaroff, Jean, and John Comaroff. 2003. "Transparent Fictions; or, the Conspiracies of a Liberal Imagination: An Afterword." In *Transparency and Conspiracy: Ethnographies of Suspicion in the New World Order*, edited by Harry G. West and Todd Sanders, 287–299. Durham: Duke University Press.

Dzingirai, Vupenya. 2003a. "'CAMPFIRE is not for Ndebele Migrants': The Impact of Excluding Outsiders from CAMPFIRE in the Zambezi Valley, Zimbabwe." *Journal of Southern African Studies* 29 (2): 445–459. doi:10.1080/03057070306208.

Dzingirai, Vupenya. 2003b. "The New Scramble for the African Countryside." *Development and Change* 34 (2): 243–264. doi:10.1111/1467-7660.00304.

Ferguson, James. 1999. *Expectations of Modernity: Myths and Meanings of Urban Life on the Zambian Cobberbelt*. Berkeley, CA: University of California Press.

Ferguson, James. 2006. *Global Shadows: Africa in the Neoliberal World Order*. Durham: Duke University Press.

Ferrier, Simon, George V. N. Powell, Karen S. Richardson, Glenn Manion, Jake M. Overton, Thomas F. Allnut, Susan E. Cameron, et al. 2004. "Mapping More of Terrestrial Biodiversity for Global Conservation Assessment." *Bioscience* 54 (12): 1101–1109. doi:10.1641/0006-3568(2004)054[1101:MMOTBF]2.0.CO;2.

Frank, David John, Ann Hironaka, and Evan Schofer. 2000. "Environmentalism as a Global Institution: A Reply to Buttel." *American Sociological Review* 65 (1): 122–127. doi:10.2307/2657293.

Frost, Peter G. H., and Ivan Bond. 2008. "The CAMPFIRE Programme in Zimbabwe: Payments for Wildlife Services." *Ecological Economics* 65 (4): 776–787. doi:10.1016/j.ecolecon.2007.09.018.

Garland, Elizabeth. 2008. "The Elephant in the Room: Confronting the Colonial Character of Wildlife Conservation in Africa." *African Studies Review* 51 (3): 51–74. doi:10.1353/arw.0.0095.

Gibson, Clark C., and Stuart A. Marks. 1995. "Transforming Rural Hunters into Conservationists: An Assessment of Community-Based Wildlife Management Programs in Africa." *World Development* 23 (6): 941–957. doi:10.1016/0305-750X(95)00025-8.

Gunderson, Lance H., and Crawford Stanley Holling, eds. 2002. *Panarchy: Understanding Transformations in Human and Natural Systems*. Washington, DC: Island Press.

International Union for the Conservation of Nature (IUCN). 1980. *The World Conservation Strategy*. Gland: IUCN.

Irwin, Alan. 1995. *Citizen Science: A Study of People, Expertise and Sustainable Development*. New York: Routledge.

Jachmann, H., and M. Billiouw. 1997. "Elephant Poaching and Law Enforcement in the Central Luangwa Valley, Zambia." *Journal of Applied Ecology* 34 (1): 233–244. doi:10.2307/2404861.

Jones, Brian, and Chris Weaver. 2009. "CBNRM in Namibia: Growth, Trends, Lessons, and Constraints." In *Evolution and Innovation in Wildlife Conservation: Parks and Game Ranches to Transfrontier Conservation Areas*, edited by Brian Child, Helen Suich, and Anna Spenceley, 223–242. London: Earthscan.

Kumar, Chetan. 2005. "Revisiting 'Community' in Community-Based Natural Resource Management." *Community Development Journal* 40 (3): 275–285. doi:10.1093/cdj/bsi036.

Leach, Melissa, Robin Mearns, and Ian Scoones. 1999. "Environmental Entitlements: Dynamics and Institutions in Community-Based Natural Resource Management." *World Development* 27 (2): 225–247. doi:10.1016/S0305-750X(98)00141-7.

Lewis, Dale M. 1989. "Zambia's Pragmatic Conservation Programme." *Pachyderm* 12: 24–26. http://www.rhinoresourcecenter.com/pdf_files/117/1175858313.pdf.

Lewis, Dale M. 1995. "Importance of GIS to Community-Based Management of Wildlife: Lessons from Zambia." *Ecological Applications* 5 (4): 861–871. doi:10.2307/2269337.

Lewis, Dale M. 2010. *Markets, Food Security and Conservation: A Model for Rural Development in Zambia*. Wildlife Conservation Society. pp. 1–11.

Lewis, Dale, Samuel D. Bell, John Fry, Kim L. Bothi, Lydiah Gatere, Makando Kabila, Mwangala Mukamba, et al. 2011. "Community Markets for Conservation (COMACO) Links Biodiversity Conservation with Sustainable Improvements in Livelihoods and Food Production." *PNSAS* 108 (34): 13957–13962. doi:10.1073/pnas.1011538108.

Logan, B. Ikubolajeh, and William G. Moseley. 2002. "The political ecology of poverty alleviation in Zimbabwe's Communal Areas Management Programme for Indigenous Resources (CAMPFIRE)." *Geoforum* 33 (1): 1–14. doi:10.1016/S0016-7185(01)00027-6.

Long, S. Andrew, ed. 2004. *Livelihoods and CBNRM in Namibia: The Findings of the WILD Project: Final Technical Report of the Wildlife Integration for Livelihood Diversification Project*. Windhoek: Directorates of Environmental Affairs and Parks and Wildlife Management, Ministry of Environment and Tourism, the Government of the Republic of Namibia. http://www.met.gov.na/programmes/wild/finalreport/TOC/TOC(p1-4).pdf.

Lubilo, Rodgers, and Brian Child. 2010. "The Rise and Fall of Community-Based Natural Resource Management in Zambia's Luangwa Valley: An Illustration of Micro- and Macro-Governance Issues." In *Community Rights, Conservation and Contested Land: The Politics of Natural Resource Governance in Africa*, edited by Fred Nelson, 202–206. New York: Earthscan.

Lyons, Andrew. 2000. "An Effective Monitoring Framework for Community Based Natural Resource Management: A Case Study of the ADMADE Program in Zambia." PhD diss., University of Florida.

Mackenzie, John M. 1988. *The Empire of Nature*. Manchester: Manchester University Press.

Marks, Stuart A. 2001. "Back to the Future: Some Unintended Consequences of Zambia's Community-Based Wildlife Program (ADMADE)." *Africa Today* 48 (1): 121–141. doi:10.1353/at.2001.0012.

Mbaiwa, Joseph E., and Amanda L. Stronza. 2011. "Changes in Resident Attitudes Towards Tourism Development and Conservation in the Okavango Delta, Botswana." *Journal of Environmental Management* 92 (8): 1950–1959. doi:10.1016/j.jenvman.2011.03.009.

Murombedzi, James C. 1999. "Devolution and Stewardship in Zimbabwe's CAMPFIRE Programme." *Journal of International Development* 11 (2): 287–293. doi:10.1002/(SICI)1099-1328(199903/04)11:2<287::AID-JID584>3.0.CO;2-M.

Musumali, Musole M., Thor S. Larsen, and Bjorn P. Kalternborn. 2007. "An Impasse in Community Based Natural Resource Management Implementation: The Case of Zambia and Botswana." *Oryx* 41 (3): 306–313.

Mutandwa, Edward, and Christopher T. Gadzirayi. 2007. "Impact of Community-Based Approaches to Wildlife Management: Case Study of the CAMPFIRE Programme in Zimbabwe." *International Journal of Sustainable Development & World Ecology* 14 (4): 336–344. doi:10.1080/13504500709469734.

Nelson, Fred, and Arun Agrawal. 2008. "Patronage or Participation? Community-Based Natural Resource Management Reform in Sub-Saharan Africa." *Development and Change* 39 (4): 557–585. doi:10.1111/j.1467-7660.2008.00496.x.

Rodrigues, Ana S. L. 2006. "Are Global Conservation Efforts Successful?" *Science* 313: 1051–1052.

Scott, James C. 1998. *Seeing like a State: How Certain Schemes to Improve the Human Condition Have Failed*. New Haven, CT: Yale University Press.

Spaargaren, Gert, Arthur P. J. Mol, and Frederick H. Buttel. 2000. "Introduction: Globalization, Modernity and the Environment." In *Environment and Global Modernity*, edited by Gert Spaargaren, Arthur P. J. Mol, and Frederick H. Buttel, 1–16. London: Sage.

Steinhart, Edward I. 2006. *Black Poachers, White Hunters: A Social History of Hunting in Colonial Kenya*. Athens, OH: Ohio University Press.

Stafford, Robert A. 1989. *Scientist of Empire: Sir Roderick Murchison, Scientific Exploration and Victorian Imperialism*. Cambridge: Cambridge University Press.

Tsing, Anna Lowenhaupt. 2003. "Agrarian Allegory and Global Futures." In *Nature in the Global South: Environmental Projects in South and Southeast Asia*, edited by Paul Greenough and Anna Lowenhaupt Tsing, 124–169. Durham: Duke University Press.

Tsing, Anna Lowenhaupt. 2005. *Friction: An Ethnography of Global Connection*. Princeton, NJ: Princeton University Press.

Twyman, Chasca. 2000. "Participatory Conservation? Community-Based Natural Resource Management in Botswana." *The Geographical Journal* 166 (4): 323–335. doi:10.1111/j.1475-4959.2000.tb00034.x.

Virtanen, Pekka. 2003. "Local Management of Global Values: Community-Based Wildlife Management in Zimbabwe and Zambia." *Society & Natural Resources: An International Journal* 16 (3): 179–190. doi:10.1080/08941920309164.

Wainwright, Carla, and Walter Wehrmeyer. 1998. "Success in Integrating Conservation and Development? A Study from Zambia." *World Development* 26 (6): 933–944. doi:10.1016/S0305-750X(98)00027-8.

West, Harry G., and Todd Sanders, eds. 2003. *Transparency and Conspiracy: Ethnographies of Suspicion in the New World Order*. Durham: Duke University Press.

Wildlife Conservation Society. 2010. "Transform a Poacher | Community Markets for Conservation." Accessed July 15, 2012. http://www.itswild.org/transform-poacher

Wolmer, William, Joseph Chaumba, and Ian Scoones. 2004. "Wildlife Management and Land Reform in South Eastern Zimbabwe: A Compatible Pairing or a Contradiction in Terms?" *Geoforum* 35 (1): 87–98. doi:10.1016/S0016-7185(03)00031-9.

Young, Zoe, George Makoni, and Sonja Boehmer-Christiansen. 2001. "Green aid in India and Zimbabwe – Conserving Whose Community?" *Geoforum* 32 (3): 299–318. doi:10.1016/S0016-7185(01)00005-7.

Zambia Wildlife Authority. 1998. *The Zambia Wildlife Act, 1998*. Lusaka: Zambia Wildlife Authority.

Zambia Wildlife Authority. 2011. "South Luangwa National Park." Accessed July 15, 2012. http://www.zawa.org.zm/index.php?option=com_content&id=23&view=article&catid=5&Itemid=30

Land beneficiaries as game farmers: conservation, land reform and the invention of the 'community game farm' in KwaZulu-Natal

Mnqobi Ngubane[a] and Shirley Brooks[b]

[a]Department of Geography, University of the Free State, Bloemfontein, South Africa; [b]Department of Geography and Environmental Studies, University of the Western Cape, Cape Town, South Africa

Scholarship on post-apartheid land reform includes research on land claims made to formal protected areas, such as national parks and state game reserves. Little attention has however, been paid to the question of land restitution claims on private lands, on which a range of nominally 'conservation-friendly' land-uses (including commercial hunting) have taken place. This article traces the emergence of the 'community game farm' as a product of land reform processes affecting freehold land in the midlands of KwaZulu-Natal province, South Africa. Two groups of land beneficiaries who were granted title to former privately owned game farms used for leisure hunting are studied in detail. The article shows that a range of state and private actors, as well as traditional authorities, have worked to ensure the continuation of the land under conservation or game farming after transfer. The central argument is that in this process, a generic narrative is imposed which works to conflate or deny the distinct historical identities of the beneficiary groups. The article raises questions about the real efficacy of land restitution in this context, as well as the appropriateness of a community-based conservation narrative when applied in the context of small farms such as those considered here.

This article considers a significant land-use change evident in the South African countryside – the conversion of private farmland to various forms of game or wildlife production.[1] While the concept of private game reserves is not new, changes in laws enabling the private ownership of wildlife have meant that, in the latter part of the twentieth century, wildlife on farms was 'transformed from a burden to an asset for landowners [resulting in] a rapid shift from livestock to game ranching across large areas of Southern Africa' (Lindsey, Romañach, and Davies-Mostert 2009, 100). The shift has taken a number of forms, from game farms offering hunting packages to generate an income; to private game reserves with upmarket tourist lodges; to property developments in the form of luxury lifestyle estates (see Brooks et al. 2011). In post-apartheid South Africa, a number of factors have fuelled this trend. They include: the growth of the tourism industry as the country reconfigured its place in the world economy; the reduced profitability of cattle farming, partly due to the removal of state subsidies for agriculture; an increase in stock theft; and landowners' reactions to new labour and land rights legislation (Carruthers 2008; Cousins, Sadler, and Evans 2008).

Game farming is being superimposed on cultural landscapes shaped by other land-uses and histories. Our geographical focus in this article is the province of KwaZulu-Natal, in particular its central interior. The area known locally as the midlands starts at Pietermaritzburg and extends north through grasslands to the 'thornveld' of the Thukela River Valley. To the north of Ladysmith is a higher-lying area of 'sourveld' (see Figure 1). Viewed as a whole, the current land tenure pattern of this region dates back to the mid-nineteenth century, when colonial-era land dispossession led to the introduction of private property and the division of territory in the Natal Colony into two main categories: so-called *native reserves*, and generous plots of freehold land to be owned by individual farmers. During the twentieth century, this landscape has been subject to various forms of displacement and racially-based dispossession, including the state's attempts to outlaw the widespread practice of labour tenancy (Surplus People Project 1983) and the forced removal of so-called *black spots*, African-owned land located in 'white' farming areas. In general, the more recent move to game farming may be viewed as a further chapter in this story of dispossession. Hart and Hunter (2004, 916) for example argue that

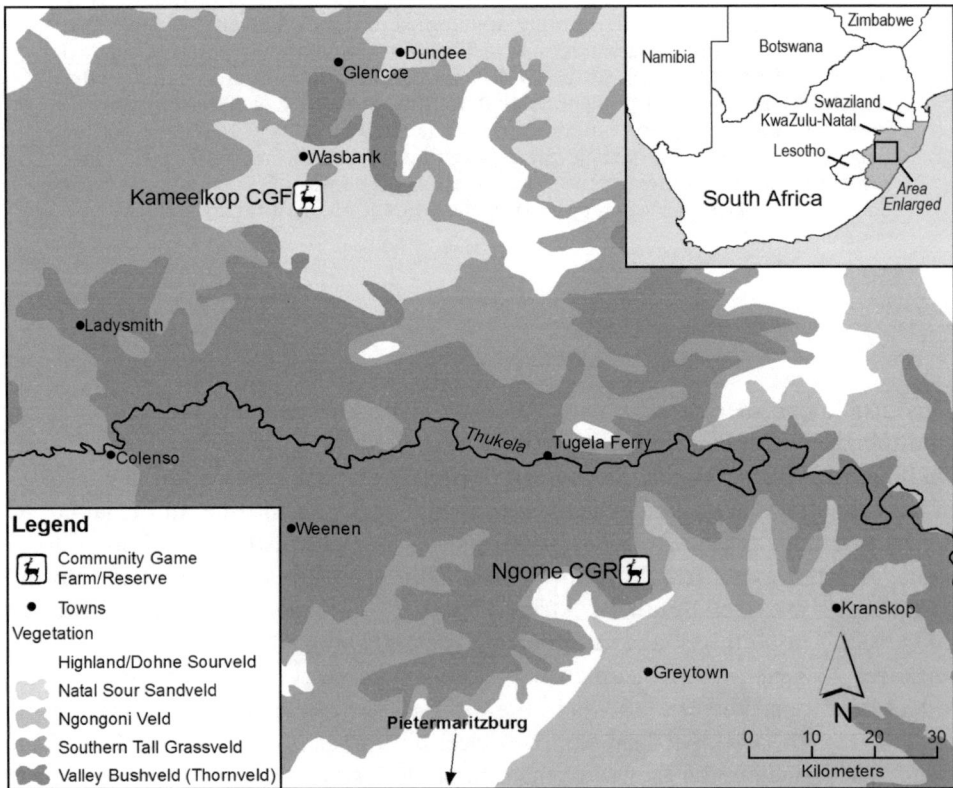

Figure 1. Map of the KwaZulu-Natal midlands with the two game farms indicated.

the 'vicissitudes of pervasive labour tenancy in this part of Natal trace an arc of dispossession from the nineteenth century to the present, as ongoing evictions of labour tenants make way for game farms'. Land rights NGOs in the region have reported an increase in the number of farm dwellers displaced by game farms (AFRA 2004).

The main focus of this article is not, however, on displaced farm dwellers. Instead we explore the recent experience of two groups of people seen as *beneficiaries* of this land-use change. These are people who were dispossessed of land during the apartheid period and who entered into the South African state's land reform process in the hope of getting their land back. In the two cases examined here, the successful land beneficiaries now find themselves in the position of being the new owners of land already converted to game farming by its previous owners. The article seeks to examine the often problematic dynamics of game farming when proffered as a solution in the context of land reform.

Game farming as an already established land-use practice on the newly acquired land is presented to land beneficiaries as, in the first place, a means of livelihood. Confronted by the evidence of a number of failed land reform projects, the state clearly hopes that these game farming enterprises will succeed where others have not. Yet game farming differs from forms of conventional agriculture in that it brings with it a set of ideas about conservation, drawn largely from the ideology and practice of community-based conservation (CBC). Game farm beneficiaries have entered into partnerships with state and especially private organisations that are able and willing to offer them wildlife management skills. Whether advising the new owners on management practice, or actually managing the game farm and associated hunting enterprises on their behalf, these partners are steeped in a conservationist discourse that reconstructs land beneficiaries as 'communities' and powerfully promotes game farming as a profitable and 'conservation-friendly' land-use option. In this process, the differing histories and relations to the land of beneficiary communities are smoothed out and disappear, and new power relations arise that call for careful interrogation.

Conservation and land claims in the South African context

To date, the literature on land claims and nature conservation in South Africa has focused on groups laying claim to state-run protected areas. The history of forced removals for conservation forms an important strand in the history of land dispossession in southern Africa as a whole (Fabricius, Koch, and Magome 2001; Ramutsindela 2003), and land claims have been seen as a strong mechanism for correcting the balance of power between communities and conservation authorities in the region (Reid 2001, 138). In the context of post-apartheid South Africa, Ramutsindela notes that claims on formal conservation land do legitimately fall under the Restitution of Land Rights Act (1994), which is 'concerned with all victims of racially motivated removals in both urban and rural areas [which includes] people who were removed from areas earmarked for national parks and nature reserves' (Ramutsindela 2003, 43).

However effective lobbying by conservationists has ensured that restituted lands are not taken out of conservation management (Ramutsindela 2002; Lahiff 2009). A cabinet memorandum has asserted, controversially, that 'conservation is a land management issue (land use) and not a land ownership issue' (AFRA 2003, 7), and land rights have been 'returned' to communities in terms of agreements that insist on the continuation of conservation management (Manjengwa 2006). The Makuleke land claim in Limpopo province, settled on 30 May 1998, was a landmark case in this regard (Ramutsindela 2002). The implications of the official policy on conservation land are summarised by Lahiff (2009, 98):

> Much of the land transferred (or 'delivered', to use the official term) under the restitution programme has been transferred in nominal ownership only, as it remains incorporated into nature reserves and state forests and, in terms of restitution agreements, is not accessible for direct use by the restored owners.

There are a number of options or models for the management of so-called *community-owned* protected areas, including co-management structures and agreements that cater for skills transfer, where claimants are considered unprepared to take on their management responsibilities (AFRA 2003). A radical question posed by scholars is whether this really constitutes restitution in the full sense of the word (Ramutsindela 2002; Walker 2008). Kepe, Wynberg, and Ellis (2003, 19) likewise point out that:

> While they may have won their land rights on paper, in practice local communities are often at the mercy of conservation agencies who tend to pursue conservation goals and the prevention of the consumptive use of natural resources [by these communities] at all costs.

Little attention has thus far been paid in the literature to the outcomes of land reform processes on conservation land *outside* of formal protected areas. In the case of land beneficiaries who have been awarded functioning private game farms as part of the restitution process, a number of themes can be identified that echo the developments outlined above. In the context of KwaZulu-Natal, the provincial conservation agency has played a strong advisory role and facilitated partnerships with private sector players able to offer wildlife management expertise. As in the case of state conservation land, a persuasive argument has been made for wildlife production or 'conservation' to remain the land-use after the land transfer has taken place. In the case of the two game farms considered here, land beneficiaries have partnered with a private sector organisation, the KwaZulu-Natal Hunting and Conservation Association (KZNHCA), which advises on wildlife management and runs the actual hunting operation.

At one level, this is comparable to other agricultural endeavours on restituted land, where those formerly in control of the land may be reconstituted as farming 'mentors' (Walker 2008). However, game farming differs significantly from more conventional farming activities because it is inextricably interwoven with conservationist narratives and ideologies. Along with these partnership arrangements has come a powerful set of ideas about conservation on so-called *community* land. These ideas come from the southern African experience in which conservationists have attempted to link livelihoods to wildlife protection through CBC initiatives (Tyman 2000; Hulme and Murphree 2001; Logan and Moseley 2002; Blaikie 2006). As in the context of state conservation areas, it is important to unpack the generic notion of 'the community' in wildlife-based land reform initiatives on private land, and to critically examine the process through which very different groups of beneficiaries are conflated into a single category. It is also appropriate to raise questions about the real efficacy of restitution for the new owners of the land. Ideas about CBC were developed in the context of extensive communal lands; these concepts may be less than appropriate to the small game farming operations conducted on farms considered here. We will argue that this (re)conceptualisation is, however, functional for stakeholders in the sector.

Conservation narratives and the role of partner organisations in 'inventing' the community game farm

Our research focused on two midlands game farms that have been transferred to land beneficiaries: Bhambatha's Kraal Private Game Reserve, renamed Ngome Community Game Reserve, near Ladysmith, and Kameelkop Game Farm (now the Kameelkop Community Game Farm), near Greytown (see Figure 1).[2] In both cases, the involvement of outside partners in the form of a provincial state department and a private sector organisation – the KZNHCA – has been influential in the process characterised here as the invention of the community game farm in the KwaZulu-Natal province. We begin by looking at the way outside 'experts' have reconceptualised these spaces and their new owners, discursively recasting them as a single generic category in a new language derived from the world of biodiversity conservation and, more specifically, that of CBC.

Ezemvelo KwaZulu-Natal Wildlife (EKZNW)

The provincial conservation authority, EKZNW, has been instrumental in promoting the continuation of game farming as a land-use by land beneficiaries. While game farms as privately owned spaces do not form part of the formal conservation estate, EKZNW is quite closely involved in private game farming due to the fact that it is tasked with overall control of wildlife and hunting in the province. No wildlife can be moved or hunted without a permit issued by the agency. In the case of the new 'community game farms', its involvement goes deeper than the usual conducting of annual inspections and the issuing of hunting permits. The provincial conservation authority has constructed its role here in a particular way. Unlike in a formal protected area, it is not directly involved in managing the transferred game farms; however the conservation authority sees itself as in some sense in partnership with the Department of Rural Development and Land Reform (formerly the Department of Land Affairs), which negotiates and oversees the land transfer itself. As an official from EKZNW put it, 'Land Affairs has given them [the beneficiaries] land. It is our role again to say that this is a biodiversity issue, and come in from that angle' (Interview with EKZNW official, July 2009).

EKZNW is interested not only in protected areas under its direct control, but also in the extensive communal lands under customary tenure, where it attempts to spread the message of biodiversity protection by drawing on the CBC narrative. This construction is readily extended to game farms transferred to land beneficiaries. One official outlined his understanding of the conservation agency's role at Kameelkop Community Game Farm. It is clear from this quotation that he views Kameelkop's new landowners in very much the same terms as so-called *surrounding communities* – that is, black communities living adjacent to state protected areas:

> In order for the conservation of biodiversity to be meaningful it has to take into account community involvement. Biodiversity conservation has to acknowledge the surrounding communities. In fact we conserve biodiversity inside and outside of protected areas, whether in Kameelkop or not. In other words, we have continuous relations with Kameelkop. Our [community] relations form a core function in our duties. We as an organization – we are there to support and advise them. (Interview with EKZNW official, August 2010)

Another form of involvement also emphasised this understanding. EKZNW invited not only the members of the new land trusts, but also chiefs or traditional authorities to attend nature conservation workshops intended to impart conservation knowledge to the new landowners. The fact that traditional authorities were often in the forefront of the land claim makes them important role players, together with the conservation agency, in the creation of the community game farm. This is illustrated in the following comment on the part of the chief who championed this claim, resulting in the acquisition of Kameelkop Game Farm:

> I had to teach them [the beneficiaries] to respect and love animals so that the future generations will also see them ... I had to teach them not to chop down trees and vegetation; the game farm is not there for firewood. Even the grass is not there for us to burn, but for the animals to graze on ... Those were amongst the things we had to do in order to train people on what they should do. I got that knowledge from ... the Nature Conservation people, who taught us about nature conservation. (Interview with Inkosi Kunene, July 2010)

In the case of Kameelkop Community Game Farm, the provincial conservation agency extended its involvement by negotiating for a former EKZNW employee to be hired by the new land Trust as manager of the game farm. As an official from the conservation authority put it, 'Land Affairs would not have known what kind of person is needed – they just provided money [to buy the land] ... our involvement comes in because it is our discipline'. The conservation agency approached the land Trust and 'told them about a retired guy who has good experience – that's why Mr X is here [as farm manager]' (Interview with EKZNW official, Kameelkop, July 2009).

At one level, this is a value-neutral development based on practical realities on the ground, in particular the fact that the land beneficiaries have not owned or managed a game farm before. At another level though, it must be recognised that Communal Property Associations (CPAs) or land trusts entering into partnerships with farming and (in this case) conservation 'experts' are also entering into a set of power relations that shape their future in particular ways. Hebinck, Fay, and Kondlo (2011) discuss the role of partner institutions in agriculture, arguing that the appointment of farm mentors who possess the requisite 'expert' knowledge usually works to ensure the continuity of land-use after land is transferred to land beneficiaries. This is quite evident in the case of game farms, with the added element of a prescriptive land-use discourse articulated around biodiversity conservation concerns.

The KwaZulu-Natal Hunting and Conservation Association (KZNHCA)

A second 'expert' organisation has also become involved with the game farms under review. While the role of EKZNW is necessarily limited to an advisory capacity, the conservation agency suggested the closer involvement of a private organisation, the KZNHCA, and effected the introductions. This private sector association is involved at both Kameelkop Community Game Farm and at Ngome Community Game Reserve, running the commercial hunting operation at both game farms in terms of signed partnership agreements. The advantage for the land Trusts is that they do not need to engage in marketing the farms so as to locate hunting clients or organise professional hunting packages; all this is done by the private sector partner from its head office in Durban.

The hunting association's narrative with regard to the 'community game farms' is one of urgently needed biodiversity education. KZNHCA is concerned to position itself as a conservation organisation, not just a body that facilitates leisure hunting for its members:

> Until now, we were essentially a hunting association, but [now] we call ourselves KZN Hunting and *Conservation*. The conservation up till now was isolated activities. Our approach is new now ... We need to manage biodiversity ... We have to look at the total environmental systems and we have to manage that. It is essential that we hunt here [at Kameelkop], because it is a fenced area and it's an isolated animal population, so there will be growth beyond the carrying capacity ... But hunting is not the first and foremost thing we want to do. For us to be able to harvest the Impala, it must have land to live on, it must have food, it must have water, it must have shelter. All those systems to make this a viable entity must be managed. We must look at the earth in its totality. We can't look at individual elements. That is what we want to teach the people. (Interview with KZNHCA official, July 2010)

The following quote from the same interview is revealing of the way the KZNHCA, like the provincial conservation authority, conflates different groups of land beneficiaries to create a single category of people – (black) 'communities' who now own land, but who do not know how to use it responsibly, and who therefore need to be taught the principles of CBC:

> The challenge that we face here is a simple one; here [at Kameelkop Community Game Farm] and in Bhambatha's [Ngome Community Game Reserve]. That is [that] the skills level of the people involved is completely inadequate to make what they got [from government] into a viable enterprise. The people are not educated from an academic perspective. The approach to what they have is still a very traditional approach. It's there to be used: you chop the tree, and you eat the meat. Which is short term. I think what we want to do essentially is to establish a longer term management approach with both the communities. (Interview with KZNHCA official, July 2010)

This construction ignores the complexity of the individual histories described in the next section, eliding them into a single narrative and in the process constructing these land beneficiaries as ignorant and in need of re-education. Such a formulation speaks, of course, to a wider dynamic present in a number of conservation contexts, not only in South Africa. It also fails to understand the significant ambiguities associated with the actual processes of restitution. These begin to emerge in the next section.

How land reform created 'community game farms' on private land in KwaZulu-Natal: Two geographies of dispossession

Partner organisations are often unaware of or uninterested in the actual identities of the people with whom they have partnered. The two groups of beneficiaries discussed here have quite different histories that stand in stark contrast to the process of generic 'community' creation outlined above. These stories are presented here in some detail, both to reveal the processes through which land reform beneficiaries are forced to take on new identities in the land reform process – in this case, identities shaped by a biodiversity conservation discourse – and to reveal how problematic and contested the beneficiaries' actual experiences of restitution can be in a context such as this.

Case study 1: from 'Bhambatha's Kraal' to the Ngome Community Game Reserve

In 2004, the Ngome Community Game Reserve near Greytown was described by the African Conservation Association in glowing terms. 'A successful land claim in the KwaZulu-Natal midlands', it claims, 'is about to become a successful hunting concession, with the potential for substantial income for about 600 previously-destitute families' (African Conservation Foundation 2004). Behind the headlines and photo opportunities, however, is a complex history of dispossession and at best partial restitution.

The farm, which is now part of the Ngome Community Game Reserve, was converted to game farming by its previous owners in 1974. Prior to this, the farm in question, Aangelegen, was a so-called *labour farm* with an absentee landowner, occupied by African families living there as labour tenants. At the time of the conversion to game farming, 15–20 labour tenant families and their cattle were evicted from the farm, leaving only three households behind in the reserve (Ngubane 2012).[3] The farm then took on its new identity as 'Bhambatha's Kraal Private Game Reserve', a name heavy with irony in this context. The labour tenants who had been evicted from the land owed their allegiance to the Zondi clan whose chief – the famous Bhambatha – unsuccessfully opposed colonial authority at the start of the twentieth century, leading to the 1906 Bhambatha Rebellion (Guy 2005). The new game farm bore Bhambatha's name, but no longer had space for Bhambatha's people.

How did the Zondi people end up as labour tenants on white-owned farms in the 'thornveld' region? These farms had their roots in the brief period of Voortrekker control of the Natal interior. The short-lived Republic of Natalia (1839–1842) made a number of land grants to trekboers who were attempting to escape British rule at the Cape. After the British took over the Natal Colony in 1843, a decision had to be made about the validity of these land grants. It was decided to formalise the grants via a 'quit rent' system: that is, the claims were recognised as valid as long as a small annual rent was paid to the state. Later, many of these farms were converted to freehold tenure (Brooks 1996). Aangelegen was converted from quit rent to freehold tenure in 1920.

The people living on Aangelegen farm – on the lands they knew collectively as Ngome – were labour tenants. This meant that they were part of a system of economic survival that became widespread in the Natal interior from the late nineteenth century. As land outside the so-called *native reserves* had been privatised, many African homestead heads reached an accommodation with landowners that allowed them to maintain their homesteads and cattle on white-owned farms in exchange for the provision of labour (McClendon 1995, 2002). McClendon (1995) explains that in terms of these verbal agreements, young men worked on the farm for part of the year where they were paid 'nominal wages, if anything'. For the rest of the year, 'the men either "rested" – that is, worked on their own homesteads on the farm – or migrated to the cities, especially Johannesburg and Durban for work at considerably higher wages' (39).

In some cases, midlands farmers purchased relatively cheap and unproductive land located in the drier north-eastern (thornveld) parts of the colony at a distance from their main commercial enterprise, both to ensure a constant supply of labour (McClendon 1995) and for winter grazing for cattle (Brooks 1996). This allowed the landowner to concentrate his main farming activities in the more productive part of the midlands (the southern part), and keep a 'spare' farm further north. According to McClendon, 'Thornveld homesteads were subject to little supervision' (1995, 52). These farms were known as 'labour farms' and Aangelegen, prior to its conversion to game farming, was such a labour farm.

The Zondis gained title to Aangelegen farm in 1997 through the post-apartheid land reform process. The final settlement did not, however, transfer land title to the actual labour tenant families evicted to make way for Bhambatha's Kraal Private Game Reserve in 1974, or even to the few households that had been allowed to stay behind as workers. This was because the land claim formed part of a larger 'tribal' land claim lodged against a series of farms by the Zondi chief, Inkosi Khulekani Zondi, on behalf of a much broader Zondi community. The Zondi chief, at the forefront of the land claim, first lodged a restitution claim but was advised by the state that this would not succeed because the original dispossession of Zondi land had taken place well before the 1913 Land Act (specified in the Restitution Act as the cut-off point for claims). The chief was advised instead to work through the land redistribution programme, a route that proved successful.

The landholder since 1997 is the Ngome Community Land Trust, a CPA set-up to hold and manage the land in the interests of the land beneficiaries. Using the pooled household grants of all the listed beneficiaries, the state was able to purchase two properties – one of them Aangelegen – for the Ngome Community Game Reserve. A former project manager at the Department of Land Affairs described the early days of the land transfer from his point of view. He understood the chief to have been influential in determining the future land-use, that is, game farming:

> [Inkosi Khulekani Zondi] knew those two land owners quite well. The relationship was complicated because people historically always believed that was their land, in which they were right. But he said: 'No, look, we'll keep this as a game reserve, let us have a positive relationship in terms of environmental conservation and also maybe we can earn an income from these farms'. So, that was the purpose of that claim, that application. (Interview with former DLA Project Officer, October 2010)

Inkosi Khulekani Zondi did not serve on the Trust because he recognised that, in terms of the legislation governing CPAs created through the land reform process, the traditional authority ought to maintain an appropriate distance from the new CPA (see Oomen 2005). His successors, however, were to be far more closely involved in decision-making around the 'community' game farm.

With regard to skills transfer, it seems that the previous owners were not approached by the Trust or by the Department of Land Affairs (DLA, now the Department of Rural Development and Land Reform) about the possibility of a joint venture. According to one of the former landowners:

> I have never been approached, nor have the other owners of the land, nor have they ever been approached to assist with the farming. We ran a safari hunting business - we had a lot of professional hunters hunting there because we had very good Inyala and we had very good Kudu and Bushbuck . . . We sold our cattle and we concentrated only on game – we had a lot of game. We were doing good business. (Interview with previous owner of Bhambatha's Kraal Private Game Reserve, November 2010)

The DLA did however appoint the previous manager of Bhambatha's Kraal Private Game Reserve to stay on and work with the new owners as a 'Training Consultant' to manage the game and the hunting camp. According to the manager, 'There was nobody there that could have dealt with the hunting, or hosted the international hunting clients' (Interview with former Training Consultant, September 2010).

At first this continuity seemed to pay off; but when problems began to arise, such as poaching and people 'settling too close to the boundary', the manager felt that the commercial viability of the game farm was under threat and he resigned in 1998 (Interview with former Training Consultant, September 2010). At this point the Trust and the Zondi chieftainship took the initiative in inviting other outside partners in to run the game farm. Successor chiefs to Inkosi Khulekani Zondi were less scrupulous about maintaining the independence of the Trust, the history of which has been marked by dissent and controversy. The traditional authority has played an increasingly dominant role in management decisions regarding the 'community game farm'. In brief, Inkosi Khulekani Zondi's immediate successor Sakhisizwe Zondi invited two new ('white') advisors on board, one of them the former Inkatha Freedom Party politician Walter Felgate. These advisers convinced the Trust that 'hunting is not productive enough. They recommended tourist attraction' (Interview with member of the dissolved Trust, June 2010). This ushered in a period of more intense development of the game farm, including the building of a hotel.

Inkosi Sakhisizwe however died and *his* successor Inkosi Mbongeleni Zondi used corruption charges against one of the advisors to ensure the dissolution of the first Ngome Community Land Trust. A new Trust was established.[4] This Trust, under the leadership of the new chief, spearheaded the construction of a new luxury lodge and conference centre to the value of R7.6 million. Bhambatha Lodge was paid for by the provincial department of Economic Development and Tourism and was officially handed over to the Ngome Community Land Trust at a ceremony in May 2009.[5]

Currently the Ngome Community Land Trust is viewed as indistinguishable from the traditional authority. Local people see the game farm more or less as the chief's private fiefdom. Several incidents are cited by interviewees to support this view. First, in consolidating its control, action was taken by the new Trust to evict the three labour tenant homesteads who had remained in the Ngome Community Game Reserve in 1974 when the other homesteads were evicted to create Bhambatha's

Kraal Private Game Reserve. Another grievance is the exclusion of commoners' cattle from the reserve:

> Most of the time [the chief] does not do things transparently – he wants to be the only one benefiting. You see, [commoners'] cattle are not allowed to graze in the game reserve – but the chief's cattle have access inside the game reserve. (Interview with Zondi community member, June 2010)

Thirdly, Inkosi Mbongeleni Zondi is regarded as the person responsible for the erection of a fence, which has extended the Ngome Community Game Reserve and restricted access to grazing land, water and firewood resources.

The change in emphasis in the period from 1997 to the present is captured in a striking passage from an interview with a member of the first Trust. He describes a transition from Ngome as a *community* game reserve to becoming a *traditional authority* game reserve:

> You see, that was not negotiated, we just saw the fence. That is because they changed the condition of the game reserve from being a community game reserve to a traditional authority game reserve. It is the traditional authority that governs here, not the beneficiaries – the owners of the land are not governing. The owners of the land are ill-treated by being oppressed by the traditional authority – that is why things are the way they are. The chairman of the Trust is involved with the traditional authority. That is why he was unable to convene a meeting with the land beneficiaries to explain what is going on. The law says the Trust must convene a meeting on a specific date and engage the community, so that the beneficiaries can express their views, in order for us to develop this [the game reserve]. That all came to an end when [the first Trust] came out of office, and the game reserve was later managed by the traditional authority. (Interview with member of the dissolved Trust, June 2010)

Unsurprisingly, there is considerable resentment on the part of many land beneficiaries towards Amakhosi [chiefs], community trusts, communal property associations and land reform in general. The labour tenant families in particular – the former occupants of the farm and the direct victims of forced removal when the private game farm was created in 1974 – feel they have received no benefit from the land reform process. They contrast their situation unfavourably with that of labour tenants on the farm located next to Aangelegen, Olivefontein. This land, which used to be part of the Bhambatha's Kraal Private Game Reserve, subsequently became a game farm in its own right named Khobotho Private Game Reserve. Unlike the Aangelegen story, the farm was claimed directly by the evictees who lodged a labour tenant claim. While the original intention was to retain game farming as the land-use after the land transfer, the Department of Rural Development and Land Reform was unwilling to purchase the game. The former owner had the wildlife shot prior to his departure. With no wildlife, and no financial basis for restocking the farm, the land has been occupied by its new owners and used for settlement and cattle keeping: Khobotho Game Reserve is no more.

Case study 2: the Boschhoek removal and Kameelkop community game farm

The second 'community game farm' studied, Kameelkop, is located near Ladysmith in the KwaZulu-Natal midlands and is run by the Boschhoek Community Trust.

The new owners of Kameelkop private game farm, the AbeKunene, have a very different history to that of the Zondi former labour tenants, both in class terms and in terms of their relationship to the land. The narrative of 'community game farming', however, works to elide these substantive differences in background and historical geography.

Unlike the Zondi labour tenants, the AbeKunene were landowners during the colonial period. The farm Boschhoek, close to Wasbank in the Ladysmith area, was bought on the open land market in the period before 1913 and was held under freehold tenure. The AbeKunene people fell victim to the apartheid state's policy of forced removal – in this case the 'clearing' of so-called *black spots* in areas designated for white farming – in 1968. The history of the AbeKunene community needs to be understood in the context of the emergence of a small peasantry, predominantly Christian or *kholwa*, within African society in the Natal and Cape Colonies from the mid-nineteenth century. 'For a brief couple of decades this group flourished and grew into a recognisable, frequently prosperous peasantry and it was from this new class of African farmers and entrepreneurs that the first African land owners came' (Surplus People Project 1983, 24; Bundy 1979).

In Natal, most land purchases by Africans were made in the interior region where land was cheaper, rather than along the coast. A group of the Kunene people, who were Swazi, moved south with their chief Sigweje and settled on mission land at Edendale outside Pietermaritzburg in the nineteenth century. The chief converted to Christianity and became a so-called *kholwa chief*. As such, he was not incorporated into the administrative machine of the Natal Native Trust – he was not regarded as a 'traditional' or 'tribal' chief as in the Zondi case – but was treated by the Natal government as a senior person in the *kholwa* community.[6] With the help of the mission, Sigweje and his people identified a farm for sale in the interior region near Ladysmith and purchased it in 1870.

Under the Nationalist government, the term 'black spots' was increasingly used to refer to pieces of black-owned freehold land, which were scheduled for removal.[7] The so-called *black spot removal* programme in Natal 'got underway seriously in northern Natal in the 1960s, with the targeting of a series of farms in the Vryheid, Newcastle and Dundee districts' (Walker 2008, 84). These removals, which often ignored title deeds held by black landowners, were not only 'bitterly opposed by the people affected' but were widely covered by the liberal press, generating negative publicity at home and condemnation abroad (Surplus People Project 1983, 102).

The AbeKunene community of Boschhoek farm was removed in 1968. An isiZulu newspaper article from the time quotes the AbeKunene chief, Inkosi Inca Kunene (the father of the current chief), as saying: 'We did not want to move. We did not approach the government and request to be removed, nor did we ask for their assistance and advice. We want to stay here, in Boschhoek' (*Ilange lase Natal* 1968).[8] The chief at first tried to resist the removal to the resettlement camp Vergelegen, some 20 km away, but ultimately saw no option but to bow to the demands of the apartheid government. This outcome was never accepted by the AbeKunene community. During the apartheid era, Inkosi Inca Kunene and prominent men in the community explored every possible avenue for the return of Boschhoek farm. Inkosi Siphiwe Kunene took over from his father in 1989, inheriting the role of leader in the struggle for the return of the lost land.

The Boschhoek Community Trust launched a land claim under the Restitution of Land Rights Act as early as 1996 but this was only finalised in 2005. As Inkosi Kunene put it:

> What we did was, we fought for the return of the land as a community. As Inkosi I had to play a vanguard role, to lodge the land claim and speak on behalf of people, based on our history, where we are originally from. I spoke on behalf of the people, based also on the Trusts, in order to ensure continuity – they know my father, grandmother and grandfather. (Interview with Inkosi Kunene, July 2010)

Two farms were awarded to the AbeKunene in the restitution agreement: Kuickvlei farm, intended for settlement purposes, and Kameelkop, a game farm about 4000 ha in size. The award of the farm Kameelkop – formerly a privately owned game farm – was made as compensatory land because the original farm, Boschhoek, is now occupied by the South African National Defence Force (SANDF). The SANDF took over the land for use as a training camp in 1972, and was not prepared to release the farm for restitution. The fact that the AbeKunene community were forced to accept compensatory land as settlement of the land claim remains controversial for some community members: the importance of the original land is suggested by the choice of the name Boschhoek Community Trust as the body to administer the restitution land awarded.

Inkosi Kunene was elected chairperson of the Boschhoek Community Trust, a position he maintains until the present day although he insists that decisions are taken in a democratic manner in line with the previous practice at Boschhoek farm. It must be emphasised that the AbeKunene people have no previous experience of game farming and indeed have not engaged in any form of farming since the forced removal, as this was impossible in the relocation camp. Most are employed in professions such as teaching, nursing and the civil service, and they work in town. Theirs is thus a very different case from the Zondi labour tenants, who were deeply affected by the establishment of the Bhambatha's Kraal Private Game Reserve when their cattle-based livelihoods were fatally disturbed by the associated removals. There was initial scepticism about the game farm, which Chief Kunene and the Trust worked hard to overcome. Inkosi Kunene played a lead role in 'converting' the community to game farming:

> Game farming was new to us, but we were very much interested [in business]...When we lodged the land claim we had agreed to make the land profitable once the land claim had been settled...When we received the game farm, we received a well organised business, and we did not have to start from scratch – erection of the boundary fence, etc. Everything was there, animals were there, and even accommodation, the lodge was there. All we had to do was to come in and take control. My role as Inkosi was to make the community understand the importance of the game farm. (Interview with Inkosi Kunene, July 2010)

However some community members have not accepted the land settlement. The question of the return of the original land remains a live, although submerged, issue at Kameelkop and may contribute to a relatively weak sense of ownership of the game farm. The game farm belongs to the AbeKunene but is located on land to which they have no prior attachment. Not only that, but there is a counterclaim by the former labour tenants of Kameelkop, who were removed when the game farm was established and who resent their land being given to other people. Some AbeKunene community members say they like the idea of owning a game farm, but they want to move the game farm to Boschhoek.

When interviewed about his involvement at Kameelkop after the handover, the previous owner of Kameelkop Game Farm expressed some regret that he was not asked to serve as a game farming mentor to the new owners of the farm (Nqabayamaswazi Previous Owner, August 2010). The issue of racial tension was explicitly raised by an AbeKunene community member as the main factor behind the withdrawal of the previous owner from the community game farm. As he put it, 'At the beginning, we operated the game farm together with the previous owner. But after some time the race issue haunted us, and so he left' (Interview with AbeKunene community member, July 2009). The Boschhoek Community Trust, it can be inferred, felt little sense of ownership over the game farm in the presence of the previous owner.

While the Department of Land Affairs has provided some post-settlement support for the Kameelkop land restitution project, in recent years this appears to have declined. According to a game farm staff member:

> We have told them, 'You Land Affairs are bad, because at the beginning when you granted us this land [the game farm], the elders said they cannot manage such a business. You then promised to come on board and assist. But now you have distanced yourselves. (Interview with Kameelkop staff member, May 2010)

The AbeKunene have made a formal complaint to the Department. For its part, the Department says it will retain a Project Officer for as long as it takes: 'When we are convinced that they can sustain themselves, we will pull out. But in my experience, I have not seen a project sustaining itself. It is highly unlikely' (Interview with Project Officer, Department of Rural Development and Land Reform, May 2010).

Power dynamics in the partnerships and the economics of 'community game farming'

The last section of the paper explores in greater depth two key aspects of the 'community game farm' as it elaborated in the context of land reform in the KwaZulu-Natal midlands. First, we look at the implications for the land beneficiaries of their partnerships with 'outside experts' and the tensions within these partnerships. A number of scholars have observed that partnerships entered into by land reform beneficiaries, whether with state or private sector partners, are about power and as a result, are often characterised by conflict (Kepe 2008; Ntshona et al. 2010). Game farming is no exception. We look at conflicts and difficulties within these relationships, and the response of partner organisations when they find themselves unable to exercise sufficient control over outcomes.

One of the main causes of tension within the partnerships is the expectation of a commercially viable and wealth-producing enterprise which has been set up in the process of 'imposing' game farming as a land-use on beneficiary communities. The second aspect, therefore, is the economic viability of game farming in this context. We noted earlier that the discourse of CBC, which aims to promote conservation-friendly behaviour by making the ownership of wildlife profitable for people living on communal land, is inappropriate in the case of small game farms, such as those discussed here. It remains unclear whether private game farming does in fact constitute an economically viable prospect for large groups of land beneficiaries: as in other areas of land reform, in all likelihood the economic expectations raised cannot be met.

An issue that has arisen at both game farms is a desire on the part of the land Trusts to extend the focus away from hunting operations oriented to local South African hunters (who constitute the membership of the partner organisation KZNHCA), to try to diversify and attract other tourist business. This is the cause of significant tension in the partnership with KZNHCA. While strongly opposing the development of the Bhambatha luxury lodge at the Ngome Community Game Reserve, the hunting association was unable to prevent the Ngome Community Land Trust from going ahead with fundraising for the lodge's construction, mainly because the chief strongly supported the development.

In thinking about this tension, it is useful to refer to the distinction Cloete, Taljaard, and Grové (2007) made between local 'biltong' or meat hunters, and international trophy hunting clients who expect a different kind of safari experience. A number of private game farms in South Africa appeal to the latter clientele, offering upmarket trophy hunting and charge in US dollars. However, as a hunting client interviewed at Ngome stated, 'Most of the guys in that Association...would be quite happy to put their tent in the bush. They don't want a luxury lodge'. Pointing to the newly constructed Bhambatha Lodge, he explained: 'They don't want to stay in a place like that lodge over there...they want to sit around a fire at night and hear the jackals and not hear a generator making electricity' (Interview with hunting client, Ngome Community Game Reserve, August 2010).

Like the Ngome Community Land Trust, the Boschhoek Community Trust too wanted to undertake projects at their game farm, which the partner organisation felt were not in its interests. A proposal by the Boschhoek Community Trust to build a conference centre on Kameelkop Community Game Farm was vetoed by KZNHCA and there is a lingering feeling of resentment about this. Some of the beneficiaries hold the view that the hunting association tries to dominate the partnership too much, hampering the Trust in its efforts to develop the property. As a staff member at Kameelkop Community Game Farm put it, 'They [KZNHCA] tend to dislike state departments that offer us some help, and claim that there is no need for such development' (Interview with Kameelkop staff member, July 2009). The hunting association, for its part, remains frustrated with the Boschhoek Community Trust, feeling that there is insufficient understanding of what is involved in running a game farm. As an official put it, 'We are still battling...to try and change their point of view with regards to how important it is to manage, utilise and sustain Kameelkop, and improve in tackling the projects *that we suggest that they must do*' (Interview with KZNHCA official, July 2010, our emphasis).

The partnership between KZNHCA and the Boschhoek Trust was tested in the winter of 2010, when Kameelkop Community Game Farm was hit by a serious fire which devastated the farm (*Northern KwaZulu-Natal Courier* 2010). Aware that the game animals would die if they did not receive supplementary feeding during this winter season, KZNHCA opted not to wait for the aid promised by various government departments (which did not in fact materialise), but took the initiative to provide supplementary feeding. The cost of feeding rose steeply, leaving the

Boschhoek Community Trust heavily indebted to the hunting association. Unfortunately, just a few weeks before the fire, the Boschhoek Community Trust had purchased cattle which they hoped to integrate with wildlife at the game farm. The KZNHCA sees the purchase of cattle as a waste of community funds that could have been utilised after a fire disaster, such as that which befell the game farm in July 2010:

> ...it's important that the money gets ploughed back to Kameelkop – the income from this place, because if you look at the books, the community bought R87 000 worth of cattle that could have been used for this problem [devastation by fire] that we have here now. I can understand if there is enough money, enough funds, they can buy cattle and obviously sell the cattle and get money from there as well. [But] Kameelkop money should be invested in Kameelkop and maybe not cattle. The focus I think in some areas is not where it should be. We have discussed this with Inkosi, he understands. (Interview with KZNHCA official, July 2010)

KZNHCA maintains a relatively close relationship with the owners of the Kameelkop Community Game Farm. However, in the case of the association's partnership with the Ngome Community Land Trust, tensions have become acute. According to KZNHCA, the problems are partly a reflection of unreasonably high expectations on the part of land beneficiaries, who expect immediate returns. They often query the money generated from the hunting operation and put pressure on the Ngome Community Land Trust. The Trust then complains to KZNHCA, and so the vicious cycle continues. As the KZNHCA official put it:

> They want money now. If the [hunter's] car leaves on Sunday, on Monday they phone me, 'where is the money?' From Bhambatha's [Ngome Community Game Reserve] there is a lot of mistrust. 'But where is the money, there must be more!' They [the members of the Trust] get pressure from the local community, that say: 'There are a lot of people driving out of here with game in their vehicles, but why don't we have a job, why is there not more game guards, why is there nobody running the lodge, why are we local people not getting the benefit? (Interview with KZNHCA official, August 2010)

This vivid description raises an important issue with regard to the game farms transferred to land beneficiaries in KwaZulu-Natal. Are they commercially viable and can they offer financial returns to beneficiary communities? Have unrealistic expectations been created? These new 'community game farms', after all, are meant to be a business in which income is generated through wildlife production and the hunting industry; biodiversity conservation concerns are secondary. Only a few members of the community can be directly employed to work in the reserves as game guards or hunting guides, and yet many hundreds of families are also intended beneficiaries.

Bothma et al. (2009, 157) estimate that 'approximately half of all South African wildlife ranches are owned on a part-time basis by professional people...and are generally unprofitable'. Thirteen years after the land transfer, the former DLA Project Officer at Ngome reflected on the fact that it was perhaps unreasonable to have expected a beneficiary community of more than five hundred people to have realised meaningful benefits from 'community game farm' land which was once a family business or even a leisure farm (Interview with former DLA Project Officer, October 2010). In the case of Aangelegen farm and its neighbour Olivefontein, the game farms had not previously been anyone's main livelihood. In the interview with the former manager of Bhambatha's Kraal Private Game Reserve, we were assured that the land had previously made significant money through its hunting operations (Interview with former Training Consultant, September 2010). This may be the case; but the former DLA Project Officer put this in context, noting that the previous owners – a successful lawyer and doctor living in Durban – were 'using them [the game farms] as a weekend thing. They were not worried about the income that they generated there'. As he admitted, 'It's a completely different thing when you've got a community that's trying to earn an income out of the same kind of operation' (Interview with former DLA Project Officer, October 2010).

This is a crucial point and one that appears to have been insufficiently thought through by the proponents of the small 'community game farms' emerging from the land reform process. Aside from issues of skills transfer and management expertise, the economic viability of small hunting farms expected to generate revenue for whole communities needs to be explored in more detail.

These frustrations have resulted in the partner organisation considering options that would enable it to gain greater de facto control of the game farm spaces. The priority from the hunting association's point of view is to find land-owning partners who can offer viable hunting packages to its members. The latter travel to a range of game farms in the province for leisure hunting purposes. From the point of view of KZNHCA, land reform carries with it the possibility that large portions of this land will not remain under game farming. The organisation's involvement with the community game farms is presented to its hunters in terms of a 'social responsibility' agenda – hunters are offered the opportunity to hunt at a community game farm and thereby make a contribution to 'community' upliftment (KwaZulu-Natal Hunting and Conservation Association 2009). However the organisation's fear that game farms previously available to its members may be 'lost' to land reform, puts the altruistic aspects of this involvement in context. Involvement in the partnerships and involvement in skills transfer would, it was hoped, enable the game farms to keep running, whilst allowing the partner organisation to exercise considerable influence over management decisions.

Following serious disagreements with some land Trusts, KZNHCA has recently taken the initiative in forming a new Trust of its own, Nemvelo.[9] In forming Nemvelo, KZNHCA appears to be positioning itself as in some sense the 'saviour' of the new category of conservation land emerging in KwaZulu-Natal, namely the game farms emerging from the land reform process. A KZNHCA official pointed out that there are now 32 beneficiated game farms in the province, constituting about 100,000 hectares of land – a substantial area. And KZNHCA alone, he claims, is paying attention to a sector that, unlike the state protected areas,

...nobody is funding...Nobody teaches them how to run it, how to generate money, how to maintain it, how to get people there. Nobody understands marketing, nobody understands finance. And let me tell you, if it stays like that, there will be an uprising. (Interview with KZNHCA official, August 2010)

It is not impossible that Nemvelo Trust may soon undertake closer management of the community game farms under study, as well as others in various parts of the province.

Conclusion

This article has provided a multi-layered analysis of the experience of post-apartheid land beneficiaries on private conservation land. In exploring the complex historical geographies of two groups of claimants who received private game farms as part of the land reform process, it aimed to highlight the way in which the new category of 'community game farm' works to refashion these complex and very different stories into a generic 'community'; a 'community' of poor black people who can then be initiated into the ideology of CBC. The evidence suggests that the narrative of biodiversity conservation and CBC is functional for various state and private sector partner organisations that have helped shape this new identity for land beneficiaries. This set of ideas, developed to improve natural resource management on communal land is not, however, usefully applied to small pieces of freehold land, such as hunting farms, previously owned by a single freehold landowner. Its application in these contexts may even be damaging.

The article has also been concerned to probe the motivations of these various 'partners' and the extent to which their involvement has shaped land reform outcomes. For the provincial conservation agency, the continuation of game farming on beneficiated land must be promoted in the face of alternative scenarios of settlement or livestock farming – neither of which supports biodiversity protection. For the hunting association, continuity with the previous land-use means that the farms remain accessible to its members, who can continue to hunt there. From the point of view of the Department of Rural Development and Land Reform, the economic returns promised by the game farming sector suggest that this is an economically viable form of production that can sustain beneficiary groups. For the land trusts and the traditional authorities who have effective control of the land, too, the 'community game farm' is a success story.

However, a number of land beneficiaries see this as a story of loss and misfortune, and they express disappointment in land reform. To paraphrase the anthropologist Greg Dening, it is useful to make a distinction here between the 'real' and the 'actual'. As Dening commented in another context, his concern was not for the obvious or apparent reality of what had taken place: he did 'not care so much about what really happened'. However, 'About what actually happened, I do [care]' (Dening 1993, 77). We know the 'real' beneficiaries in this story – this article has described their specific histories of dispossession, and the transfer of land through the state land reform process is a documented matter of public record. The question we are posing here is a less obvious and more provocative one: who are the *actual* beneficiaries of the 'community game farm'?

Acknowledgements

The authors gratefully acknowledge financial support from the University of the Free State Research Cluster, 'New Frontiers in Poverty Reduction and Sustainable Development', as well as the personal support of the Cluster director, Doreen Atkinson. The authors also benefited from participation in two interrelated collaborative research projects: 'Farm Dwellers, the Forgotten People? Consequences of Conversions to Private Wildlife Production in KwaZulu-Natal', funded by the South Africa Netherlands Research Programme on Alternatives in Development; and the extended version of that project, 'Farm Dwellers, the Forgotten People? Consequences of Conversions to Private Wildlife Production in KwaZulu-Natal and the Eastern Cape', funded by NWO-WOTRO (NWO is the Netherlands Organization for

Scientific Research, partnered with WOTRO Science for Global Development in the NWO-WOTRO research programme). We are most grateful to all the research participants in KwaZulu-Natal. We would also like to acknowledge the input and insightful comments of several careful readers: Gillian Hart of the University of California at Berkeley, Thembela Kepe of the University of Toronto, Lindokhuhle Khumalo of the University of the Western Cape, Jenny Josefsson of the University of the Free State, Robin Palmer of Rhodes University, Robert Gordon of the University of Vermont (currently based at the University of the Free State), and Marja Spierenburg of the Vrije Universiteit, Amsterdam. Any errors of fact or interpretation are those of the authors. Thanks to Frank Sokolic for drawing the map.

Notes

1. These developments are significant in terms of land area, as well as economic turnover. For facts and figures on the game farming industry in South Africa [albeit some years old, see NAMC (2006)].
2. The research methodology involved close engagement over several months at the two field sites, as well as targeted interviews with present and past role-players in the story. An initial period of fieldwork was conducted in July 2009 and this was extended with longer periods of fieldwork at both research sites during 2010. All interviews were conducted by Mnqobi Ngubane. Interviews were conducted either in English or isiZulu, as appropriate. All translations from the Zulu are done by Mnqobi Ngubane. For a detailed discussion of the research methodology, see the Masters dissertation in which these findings were first presented (Ngubane 2012). The research was carried out at the University of the Free State under the supervision of Shirley Brooks, now based at the University of the Western Cape.
3. The authors have taken the decision not to provide a detailed citation linking each piece of research information to the unpublished Masters dissertation where they were first presented (Ngubane 2012). It was felt that this would be too cumbersome and interrupt the flow of the article. The dissertation is however available for consultation.
4. For a recent in-depth consideration of similar disputes over land trusts to whom (in this case, state) conservation areas have been transferred – albeit in other part of the country – see Fay (2013).
5. The construction of the Bhambatha Lodge went ahead against the advice of the partner organisation, the KwaZulu-Natal Hunters and Conservation Association. Its construction caused tensions within the partnership, discussed in the last section of the article.
6. Thanks to Jeff Guy for this insight into the historical figure of the kholwa chief.
7. Bill Freund put this well almost 30 years ago when apartheid-era removals were still ongoing: 'Black spots are, as the name implies, islands of black tenure in supposedly white zones. They have in general belonged to the more prosperous strata of the African peasantry who had been able, when it was legal before the land division of 1913, to purchase freehold property, often through companies of ex-wage workers or the agency of the missions' (Freund 1984, 51).
8. Translation from isiZulu by Mnqobi Ngubane.
9. Like the provincial conservation agency's name (Ezemvelo KZN Wildlife), this name is also derived from the Zulu word for 'nature'.

Note on contributors

Mnqobi Ngubane recently obtained his Masters degree *cum laude* in Geography at the University of the Free State. He is currently a PhD intern at the Institute for Poverty, Land and Agrarian Studies at the University of the Western Cape. His research interests include land reform, livelihoods and the exercise of power in rural contexts.

Shirley Brooks is a Human Geographer whose research and teaching has focused on environmental history, critical conservation and land issues. She is currently Associate Professor at the Department of Geography and Environmental Studies at the University of the Western Cape, having worked previously at the Universities of the Free State and

KwaZulu-Natal. Shirley has written extensively on the history of protected areas. Her current research focuses on the socio-spatial impacts of private wildlife production or game farming.

References

African Conservation Foundation. 2004. *First SA Land Claim Set Aside for Conservation.* Accessed May 25, 2012. http://mail.africanconservation.org/wildlife-news/item/first-sa-land-claim-set-aside-for-conservation

Association for Rural Advancement (AFRA). 2003. *Investigation of the Effects of Conservation and Tourism on Land Tenure and Ownership Patterns in KwaZulu-Natal.* Unpublished Report, Phase 1, September 15. Pietermaritzburg: McIntosh Xaba & Associates.

Association for Rural Advancement (AFRA). 2004. *Communities, Conservation, Eco-tourism and Tenure Security Workshop Report.* Accessed November 12, 2010. http://www.afra.co.za/upload/files/AP19.pdf

Blaikie, P. 2006. "Is Small Really Beautiful? Community-Based Natural Resource Management in Malawi and Botswana." *World Development* 34 (11): 1942–1957. doi:10.1016/j.worlddev.2005.11.023

Bothma, J., P. du, H. Suich, and A. Spenceley. 2009. "Extensive Wildlife Production on Private Land in South Africa." In *Evolution and Innovation in Wildlife Conservation: Parks and Game Ranches to Transfrontier Conservation Areas,* edited by B. Child, H. Suich and A. Spenceley, 147–162. London: Earthscan.

Brooks, S. 1996. *An Historical Overview of the KwaZulu-Natal Pilot Land Reform District, 1800–1996. Unpublished Status Quo Report, KwaZulu-Natal Pilot Land Reform Programme.* Pietermaritzburg: Department of Land Affairs.

Brooks, S., M. Spierenburg, L. van Brakel, A. Kolk, and K. B. Lukhozi. 2011. "Creating a Commodified Wilderness: Tourism, Private Game Farming, and 'Third Nature' Landscapes in KwaZulu-Natal." *Tijdschrift Voor Economische en Sociale Geografie [Journal of Economic and Social Geography]* 102 (3): 260–274. doi:10.1111/j.1467-9663.2011.00662.x

Bundy, C. 1979. *The Rise and Fall of the South African Peasantry.* London: Heinemann.

Carruthers, J. 2008. "'Wilding the Farm or Farming the Wild?' The Evolution of Scientific Game Ranching in South Africa from the 1960s to the Present." *Transactions of the Royal Society of South Africa* 63 (2): 160–181. doi:10.1080/00359190809519220

Cloete, P. C., P. R. Taljaard, and B. Grové. 2007. "A Comparative Economic Case Study of Switching from Cattle Farming to Game Ranching in the Northern Cape Province." *South African Journal of Wildlife Research* 37 (1): 71–78. doi:10.3957/0379-4369-37.1.71

Cousins, J. A., J. P. Sadler, and J. Evans. 2008. "Exploring the Role of Private Wildlife Ranching as a Conservation Tool in South Africa: Stakeholder Perspectives." *Ecology and Society* 13 (2): 43. http://www.ecologyandsociety.org/vol13/iss2/art43/

Dening, G. 1993. "The Theatricality of History Making and the Paradoxes of Acting." *Cultural Anthropology* 8 (1): 73–95. doi:10.1525/can.1993.8.1.02a00040

Fabricius, C., E. Koch, and H. Magome. 2001. "Towards Strengthening Collaborative Ecosystem Management: Lessons from Environmental Conflict and Political Change in Southern Africa." *Journal of the Royal Society of New Zealand* 31 (4): 831–844. doi:10.1080/03014223.2001.9517679

Fay, D. 2013. "Neoliberal Conservation and the Potential for Lawfare: New Legal Entities and the Political Ecology of Litigation at Dwesa-Cwebe, South Africa." *Geoforum* 44: 170–181. doi:10.1016/j.geoforum.2012.09.012

Freund, B. 1984. "Forced Resettlement and the Political Economy of South Africa." *Review of African Political Economy* 11 (29): 49–63. doi:10.1080/03056248408703567

Guy, J. 2005. *The Maphumulo Uprising: War, Law and Ritual in the Zulu Rebellion.* Pietermaritzburg: University of KwaZulu-Natal Press.

Hart, G., and M. Hunter. 2004. Women and Juniors as Historical Agents: Natal 1920s–1940s. Review of Genders and Generations Apart: Labor Tenants and Customary Law in Segregation-Era South Africa, 1920s to 1940s, by T.V. McClendon. *Journal of Southern African Studies* 30 (4): 915–917. http://web.ebscohost.com.ezproxy.uwc.ac.za/ehost/pdf viewer/pdfviewer?sid=4271133d-b6f6-4db3-9618-bbee6ee5be2d%40sessionmgr110 ≈vid=4& hid=122

Hebinck, P., D. Fay, and K. Kondlo. 2011. "Land and Agrarian Reform in South Africa's Eastern Cape Province: Caught by Continuities." *Journal of Agrarian Change* 11 (2): 220–240. doi:10.1111/j.1471-0366.2010.00297.x

Hulme, D., and M. Murphree, eds. 2001. *African Wildlife and Livelihoods: The Promise and Performance of Community Conservation*. Oxford: James Currey.

Ilange lase Natal. 1968. "Sithuthiwe isizwe sikaKunene [Kunene's people relocated]." *Ilanga Newspaper* August 24. Accessed in Killie Campbell Africana Library Collections, Durban.

Kepe, T. 2008. "Land Claims and Comanagement of Protected Areas in South Africa: Exploring the Challenges." *Environmental Management* 41 (3): 311–321. doi:10.1007/s00267-007-9034-x

Kepe, T., R. Wynberg, and W. Ellis. 2003. *Land Reform and Biodiversity Conservation in South Africa: Complementary or in Conflict? No. 25 in Occasional Paper Series, Land Reform and Agrarian Change in Southern Africa*. Cape Town: Programme for Land and Agrarian Studies, University of the Western Cape. http://www.plaas.org.za/sites/default/files/publications-pdf/OP%2025.pdf

KwaZulu-Natal Hunting and Conservation Association. 2009. "Social Responsibility." Accessed March 1, 2009. http://www.kznhunters.co.za/site/awdep.asp?depnum=29367

Lahiff, E. 2009. "With What Land Rights? Tenure Arrangements and Support." In *Another Countryside? Policy Options for Land and Agrarian Reform in South Africa*, edited by R. Hall, 92–117. Cape Town: Institute for Poverty, Land and Agrarian Studies, University of the Western Cape.

Lindsey, P. A., S. S. Romañach, and H. T. Davies-Mostert. 2009. "The Importance of Conservancies for Enhancing the Value of Game Ranch Land for Large Mammal Conservation in Southern Africa." *Journal of Zoology* 277 (2): 99–105. doi:10.1111/j.1469-7998.2008.00529.x

Logan, B. I., and W. G. Moseley. 2002. "The Political Ecology of Poverty Alleviation in Zimbabwe's Communal Areas Management Programme for Indigenous Resources (CAMPFIRE)." *Geoforum* 33 (1): 1–14. doi:10.1016/S0016-7185(01)00027-6

Manjengwa, J. 2006. *Natural Resource Management and Land Reform in Southern Africa. No. 15 in Occasional Paper Series, Commons Southern Africa*. Cape Town: Centre for Applied Social Sciences and Programme for Poverty, Land and Agrarian Studies, University of the Western Cape.

McClendon, T. V. 1995. "Genders and Generations Apart: Labour Tenants, Law and Domestic Struggle in Natal, South Africa, 1918–1944." PhD diss., Stanford University.

McClendon, T. V. 2002. *Genders and Generations Apart: Labor Tenants and Customary Law in Segregation-Era South Africa, 1920s to 1940s*. Portsmouth: Heinemann; Oxford: James Currey.

National Agricultural Marketing Council (NAMC). 2006. *Report on the Investigation to Identify Problems for Sustainable Growth and Development in South African Wildlife Ranching*. Pretoria: NAMC. http://www.namc.co.za/dnn/LinkClick.aspx?fileticket=HS7BSMTKAZY%3D&tabid=72&mid=542

Northern KwaZulu-Natal Courier. 2010. "Devastation: Animals Left Stricken after Ravaging Fire." *Northern KwaZulu-Natal Courier* July 23.

Ngubane, M. 2012. "Land Beneficiaries as Game Farmers in the 'New' South Africa: Land Reform in Relation to Conservation, the Hunting Industry and Chiefly Authority in KwaZulu-Natal." MA diss., University of the Free State.

Ntshona, Z., M. Kraai, T. Kepe, and P. Saliwa. 2010. "From Land Rights to Environmental Entitlements: Community Discontent in the 'Successful' Dwesa-Cwebe Land Claim in South Africa." *Development Southern Africa* 27 (3): 353–361. doi:10.1080/0376835X.2010.498942

Oomen, B. 2005. *Chiefs in South Africa: Law, Power and Culture in the Post-Apartheid Era*. Oxford: James Currey; Pietermaritzburg: University KwaZulu-Natal Press.

Ramutsindela, M. 2003. "Land Reform in South Africa's National Parks: A Catalyst for the Human-Nature Nexus." *Land Use Policy* 20 (1): 41–49. doi:10.1016/S0264-8377(02)00054-6

Ramutsindela, M. F. 2002. "The Perfect Way to Ending a Painful Past? Makuleke Land Deal in South Africa." *Geoforum* 33 (1): 15–24. doi:10.1016/S0016-7185(01)00008-2

Reid, H. 2001. "Contractual National Parks and the Makuleke Community." *Human Ecology* 29 (2): 135–155. doi:10.1023/A:1011072213331

Surplus People Project. 1983. *Forced Removals in South Africa, Volume 4: Natal.* Cape Town: Surplus People Project.

Tyman, C. 2000. "Participatory Conservation? Community-Based Natural Resource Management in Botswana." *The Geographical Journal* 166 (4): 323–335. doi:10.1111/j.1475-4959.2000.tb00034.x.

Walker, C. 2008. *Landmarked: Land Claims and Land Restitution in South Africa.* Athens, OH: Ohio University Press; Johannesburg: Jacana.

Constructing walls of carbon – the complexities of community, carbon sequestration and protected areas in Uganda

Adrian Nel and Douglas Hill

Department of Geography, Otago Univeristy, Dunedin, New Zealand

Carbon forestry represents a degree of continuity and discontinuity with traditional conservation practices, rescripting forestry management/governance and land access through projects on the ground in variegated, context-dependent ways. Utilising the comparative lens of two distinct projects operating on state-led protected areas in the east of Uganda, and focusing on their contested boundaries, this paper reflects on these dynamics and tries to make sense of the implications for the rural communities within the project vicinities. The projects and their framings reassert the claims to territory of the state in different ways which are contingent upon and emergent from the local institutional and historical context, or 'legacies of the land', which can be seen in context to be disputed and contested. Whilst it must be said that there can be selectively progressive elements within carbon forestry initiatives, it can be observed that techno-centric interventions, which depoliticise their local contexts and selectively transnationalise access to land and forestry resources, can further marginalise local communities in the process.

Controversies involving allegations of 'land grabbing' and enforced, uncompensated evictions have placed the spotlight on carbon forestry projects and protected areas in East Africa. These have included New Forests (UK) at Namwasa Forest Reserve in Uganda and a pilot project – Reducing the Effect of Deforestation and Degradation (REDD+) – in the Rifuji delta of Tanzania that was mooted to involve WWF. Paralleling a strong focus on 'land grabbing' in contemporary discourse on the agricultural sector in Africa, there are strains of critical geographic investigations exploring variations of 'accumulation by dispossession' (Harvey 2009), these allegations extend the debate into the sphere of carbon and forestry through what Bumpus and Liverman (2008) conceptualise as 'accumulation by decarbonisation'. These claims contradict the assumption of 'win–win' for community and environment in carbon forestry and centre concern around carbon forestry's potential impact on rural African communities. REDD+ and 'forest degradation' in Africa must be considered alongside its high population growth, poverty and contested land politics that are all so intimately tied up with what Agrawal and Narain (1991) call 'survival emissions': i.e. emissions resulting from basic needs, including agricultural practices.

These concerns pose questions about the contested nature of the spatial (re-)organisations engaged under carbon forestry projects and the way they relate to protected 'natures'. Central to this are the particular changing dynamics, roles and claims to (often disputed) territory that arise in the shifting relationship between the state, private/NGO actors and communities within project vicinities. Here the carbon economy and the processes of 'carbon fixation' are rearranging institutional structures, interacting with differing national frameworks in variegated, context-dependent ways, and with new actors and practices evident in the reproduction of new 'local' natures under the organising principle of carbon.

A range of authors have attempted to understand processes of neoliberalisation and how they relate to 'the environment' in its differentiated contexts (Harvey 2009; Castree 2003, 2008; McCarthy and Prudham 2004; Heynen 2007; Peluso 2007; Brenner, Peck, and Theodore 2010; Brockington and Igoe 2007). This 'neoliberalisation of nature' for Castree (2008), is simultaneously a social, environmental and global project, involving the renegotiation of the boundaries between the market, state and civil society so that more areas of people's lives and the biophysical world are governed by the economic logic of the 'invisible hand' which is spatially expansive and socially and environmentally exorbitant (Castree 2008).[1] At the multilateral level, for instance, Goldman (2004) analyses how 'Nature' has increasingly been incorporated into strategies of actors such as the World Bank. In so doing, he examines how these processes provide new cultural/scientific logics and domains of political–economic calculation, which reinterpret qualities of a state's territory, legitimating 'devices' for enclosing/transnationalising access to territories and 'improving' conditions of nature and populations in 'the degraded periphery' (Goldman 2004).

Carbon forestry projects can be understood to be one of this suite of broader strategies deployed to render climate change governable (Oels 2005). The main proponent of REDD+ activities and 'readiness', for instance, is the Forest Carbon Partnership Facility, housed within the World Bank. These forms of new, program-matised eco-governmentality (Agrawal 2005) find their operationalisation not only through multilateral and state institutions but also through non-state (NGO) and private entities, including private carbon providers and multinational companies. The resulting renegotiations between these entities alter or augment local governance structures to facilitate and support the framework of carbon offsetting and devolution of management to private or NGO actors. As Sassen (2005) asserts, 'globalisation' or transnationalisation can be said to unfold and be constituted locally. This also suggests we need to understand states in more nuanced ways: as complex, shifting and contingent assemblages of institutions, actors and policies enabling and also limiting access to resources (Bakker 2005). This is occurring as a process of self-maintenance and legitimation; here in the face of both the challenges and shifting forms of donor funding attendant to climate change.

An important component of this rescaling of the state is a reinforced emphasis on 'community-based conservation', in this case through the creation of 'local' sequestration sites. Indeed, this forms a core component of REDD+ and carbon forestry, privileged as the equity pillar in 'win–win' roll out, for climate and community. However, a critique this approach explores is how this devolution of management to the level of the 'local' also purports to entail community participation and ownership (Murphree 2001), the lack of which is often responsible for the failure of such programmes. Further criticisms stem from the limited degree of decision-making autonomy allowed (Tipa and Welch 2006), since states typically retain legal control over natural resources (Ribot 2002), or the lion's share of power and revenues from natural resources that they loath to relinquish to local jurisdictions (Alcorn, Kajuni, and Winterbottom 2002). Thus, while the state's roles in management entail a renegotiation under carbon forestry, changing form, scale and type of practice, the state does not 'disappear' and remains constitutive of neoliberal regimes (Peluso 2007). Enclosures of 'natures' and their assembled renegotiations of state and private do not mean that neoliberal natures are disembedded from states, power structures and existent social relations and inequalities, contra to the trope and central myth of orthodox neoliberalism of 'free (carbon) markets' (Peluso 2007). As Ferguson (2004) has argued, however, modernising initiatives can often obfuscate these realities for the purposes of implementation.

This paper focuses on two particular carbon forestry projects on protected areas in Eastern Uganda which demonstrate these issues at a variety of scales. Drawing from a Ph.D. research conducted by the first author in Uganda in 2012, this paper focuses on the boundaries of the areas and how they are interpreted and reasserted in the projects' framings in order to carve out spaces of carbon sequestration. This sets the scene for a deeper exploration of 'what it is these projects do' as spatialising entities. As we will demonstrate, the two initiatives involve interactions of state and non-state actors in rearrangements in the framing of their localities and the protected areas they take up for their selective new purposes of carbon sequestration. These have distinct spatial and social impacts – selectively re-emphasising 'fortress conservation' and unevenly re-emphasising the relationships between state and 'community'. Exploring the two carbon forestry projects in this way, this paper aims to repoliticise their respective contexts and contested legacies of the lands within which the projects become enwrapped. Finally before proceeding, it is important to keep in mind that carbon forestry and its variants are still emerging. In this vein, this paper does not seek to provide a uniform explanation for all these processes (which are diverse and sometimes contradictory), but does suggest that these overarching logics, interconnectivities and commonalities need to be comprehended if local variations are to be understood in all their complexity in different social and geographic contexts.

Project framings and community contestations

In keeping with the theme of 'old land new practices', we proceed to set out and explore the spatialising dynamics within carbon forestry as project and practice through the lens of two examples in Uganda (see Figure 1). The first is the Busoga Forestry Company (henceforth BFC), which is a subsidiary of Norwegian Multi-national Green Resources AS. Its activities in Bukaleba Central Forest Reserve involve a Voluntary Carbon Standard (VCS) – a carbon sequestration project – in Mayuge District. The second project is the Mt Elgon Regional Ecosystems Conservation Project (MERECP) on the Kenya border. The Mt Elgon Conservation project is working towards piloting a Reducing the Effects of Deforestation and

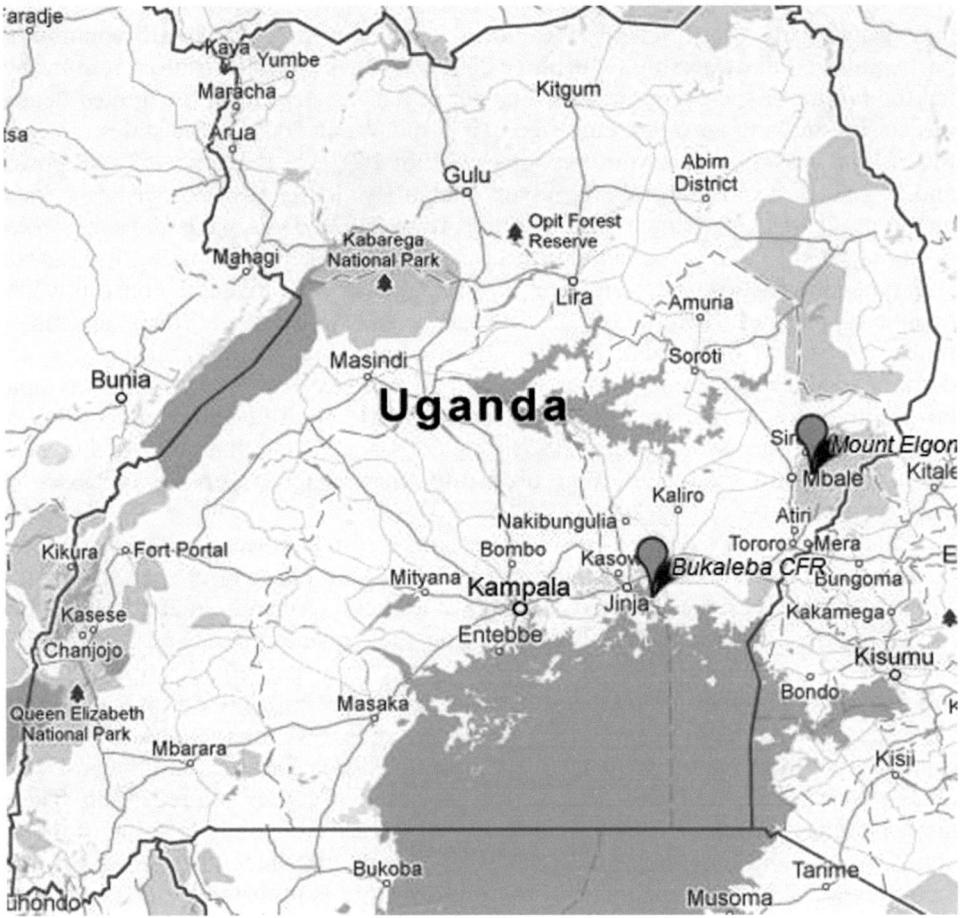

Figure 1. Project Locations.

Degradation (commonly referred to as REDD+) demonstration project to protect the sections of the Mt Elgon National Park. Both these projects are on 'protected' land, managed by the National Forestry Authority (NFA) and Uganda Wildlife Authority (UWA), respectively.

Both projects are playing out in a complex and debilitated forestry and environmental 'sector' in Uganda. This is characterised by rapid deforestation of 1.9% annually, leading to a loss of 90,000 ha of forests per year (NFA 2009). There are also increasing land pressures due to population increases, since Uganda has one of the highest fertility rates in Africa at 3.5%, as well as frequent instances of forced migration. This non-exhaustive list of pressures also involves a complicated politics of land contestation on protected areas, which is undermined by vote-seeking to bolster waning support for the ruling National Resistance Movement (NRM). Further dynamics include disruptions and inadequacies in management approaches from the colonial era compounded by the proceeding political turmoil over successive regimes. These legacies are coupled with a lack of coordination, funding and acknowledgement of conservation and environmental governance. In turn, these under-acknowledged governance issues sit in tension with the country's developmental orientation, and subsequent donor-funded reforms, which prioritised commercial forestry and what became corrupted centralised National Forestry Authority control within forestry management. Variously these reasons have led to the compromise of protected area territories and the increase in encroachment from 180,000 individuals in 2005 to 270,000 in 2009 (NFA 2009).

The legacy of these factors is evident in the fragmented physical and social landscapes at both sites, with heavy deforestation in Mayuge and in particular at the Bukaleba's compatriot South Busoga Central Forest Reserve. Bordering the Mt Elgon National Park, unsustainable land use practices and population pressures have seen the degradation of the ecosystem, with deforestation, a loss of soils due to run off and erosion. Communities are moving higher up the steep slopes of Mt Elgon to cultivate because of land competition and scarcity, which contributed to landslides in 2010 and 2012 in Bududa district and Wanale sub-county.

Moreover, both projects exhibit contested land politics that have important lessons about the types of orientation and problem framing evident within carbon forestry that we will now begin to set out. On the one hand, the BFC project frames its activities under a strict carbon approach under the VCS, which depoliticises it local context and de-legitimises alternative land claims to its project site *qua* 'old land'. This paper argues that the reason for this is largely due to the fact that the project is enwrapped in an assemblage prioritising logging interests and transnational capital. The discourse of 'encroachment' that results disregards local claims with important ramifications for local communities. In contrast, and more positively, the MERECP project unofficially acts as a buffer between the community and the Mt Elgon National Park protected area, working to mediate conflict and establish incentives for its conservation. However, its benefits are selective and limited in scope, and contoured by local power relationships, and its mediation of conflict is as much dependent on local circumstances and the agency of pre-existing community organisation as on its own activities.

The BFC's VCS project

The carbon offset component of the BFC activities at Bukaleba Central Forest Reserve, bordering Lake Victoria, comprises of 2134 ha under the VCS standard. It integrates closely into their commercial activities as an industrial plantation and timber company as a subsidiary of Green Resources AS (Norway). Green Resources' strategic orientation and activities in carbon offsets, forest products and renewable energy, are augmented by local acquisitions, including Lango Forestry and BFC in Uganda, which merged in 2012. These have seen it develop into the largest forestry operation in Africa, with 14,000 ha under production and 610,000 ha under development, of which 12,000 ha is in East Africa (with 3500 employees), with significant areas in Tanzania (34,000 ha of land, with a further 120,000 ha in the process of acquisition), Mozambique (172,000 ha) and Sudan (179,000 ha) (Green Resources 2010). It also owns East Africa's largest sawmill, Sao Hill, which was originally built by Norwegian development aid in the mid-1970s.

The BFC bought out the 50-year lease to the approx. 9500 ha Bukaleba Forest Reserve from Tree Farms in 2006. The reserve is officially the jurisdiction of the NFA and includes defined areas with specific characteristics of unplantable, conservation and riparian buffer zones, as well as a disputed area set aside for beef ranching during Amin's regime (1971–1979) that now houses a school and Christian NGO. In total, this means that approximately 5000 ha of the 9500 ha is considered plantable land suitable for forestry. The Reserve also includes an area of 500 ha set aside by Tree Farms, under duress after contestations with communities. This 500 ha was promised to community residents for forest-friendly activities in 1993 by the then Minister for Water, Lands and Environment Hon. Ruhakana Rugunda. It was subsequently claimed as a community development initiative by the BFC after its lease of the reserve. The land referred to as 'community forest management' land by the NFA was meant to be allocated to communities, as they and the BFC saw it, for the assisted planting of tree species to maintain the integrity of the Reserve. However, as the NFA range manager attests 'the interest of the community is not in growing trees, but in settlement and farming in the reserve' (NFA Official, interview by the first author, Jinja, July 2012).

By November 2011, 4134 ha of the remaining plantable 4500 ha was available for 'productive use' according to the BFC Range Manager at Bukaleba, with the overall exotic species planted at about 40% eucalyptus and 60% pine. The specific 50-year time frame of BFC's lease would afford the company two rotations of planting and harvesting from the soil which would comprise about 40–45 years of the 50-year lease. The company, as 'still a young plantation' (BFC Official, interview with the First Author, Jinja, July 2012) has no plans in place yet for its reclamation after that time. This is a cause of concern that the local District Environment Officer (DEO) emphasised in an interview, relating to the detrimental effects of the plantation on the lake and on soil fertility in the reserve. The strict focus on maximising carbon sequestration from its VCS approval in 2012 to cover trees planted after 2004 – for which it is now looking for a buyer for an expected first payment of USD 15,000 – has implications for the communities in the project areas, who are framed as 'encroachers' and which the implementers actively seek to exclude.

The company paints itself as oriented towards 'sustainable development', citing the building of roads, clinic support and employment as community benefits, the allocation of the 500 ha provided by Tree Farms as indicative of its commitment to community benefits. However, these benefits remain scant and difficult to access for the majority of people in the area. Furthermore, the 500 ha remains contested and has not effectively been allocated as official Community Forestry Management land due to a stand-off between communities and the company which holds that the land should be planted with trees not crops. This is not least because of subsequent promises to de-gazette the area at election time by politicians, including Uganda President Museveni himself in 2010 (URN 2011). Although these have not yet been forthcoming, they are 'in process', according to Idi Isabirye, the National Resistance Movement MP for Bunya South Constituency (interview with the first author, Kampala, July 2012). Communities, however, also interpreted this directive as a mandate to go into the forest reserve and claim land, with residents burning and clearing pine trees on 70 acres. Residents also destroyed another 30 acres of eucalyptus trees in one night next to the 500 ha before being stopped by the BFC and armed personnel in early 2011. As a result of this incident, the BFC served an intention to sue the Attorney General, the NFA and the District local Government over the 'failure to advise the government correctly' on the status of the area (Anti-Corruption Coalition Uganda 2010). The cropping is not limited to the disputed

500 ha however, and even at a glance the hillsides (considered unfit for tree planting) above the tree lines and below the infertile ridge tops can be seen to be filled with local maize crops.

The BFC's approach is to simplistically lay the problem at the feet of the Forestry Authority and local politicians, who either through action or inaction undermine the territorial integrity of the Reserve stipulated in the lease contract they negotiated. The Bukaleba project was unable to secure either A/R CDM certification or to attain the Climate, Conservation and Biodiversity Alliance (CCBA) standard. According to interview respondents, this failure is attributable to the issues of encroachment and a lack of demonstrable net social benefit (as indicated in the projects VCS documentation (DNV 2012)). On the ground, the company has, by its own admission, planted the 'easier' areas of the Reserve first, including most of its southern section, and the remaining areas to the east by observation are the ones most 'encroached' by communities within and adjacent to the reserve. Representatives from village local councils interviewed by the first author estimated populations to be roughly 4778 people in Walumbe and approximately 3200 in Nakalaanga both of which are in the reserve's buffer zone between the designated 'forest land' and the lake. Smaller numbers of people are also to be found at the Bukaleba village at the centre of the BC Forest Reserve. In Lwanika village adjacent to the Forest Reserve, numbers are estimated at 7000. Given the degree of contestation so far and the population numbers in the project vicinity, it remains to be seen exactly how further planting will proceed.

The company employs a full-time team of approximately 30–40 for forest protection to protect the plantation from fire, encroachment and property thefts. According to its range manager (Interview with the first author, Bukaleba, June 2012), the company has taken a 'softly softly' approach to 'encroachers'. He argues that 'the principle has not been to fight people, just to plant', protecting their trees and incrementally clearing and planting areas to free them up from encroachment.

Communities adjacent to and within the Reserve, however, consider the current situation to be extremely difficult, with few opportunities outside agriculture, declining fish stocks and rising fuel costs. Furthermore, there is a fairly pervasive negative perception of the company's impact amongst community members interviewed, even on the part of temporary employees (whose employment has dwindled now that the bulk of clearing and planting have been done) who cite low pay and poor working conditions. In the worst case, a Nakalaanga resident cited low (and irregular) payments of 50,000 Ugandan shillings a month (US$23), unpaid over 10 months. Residents decry the cutting of their crops in the designated reserve lands, although some Bukaleba village residents asserted that they had paid BFC guards who used a 'soft touch', in allowing them to do their planting. Residents also lament the lack of opportunities to plant according to the old Taungya system, where crops could initially be planted amongst the growing trees before the canopy shaded them out. Additionally, they cite the small and inadequate benefits from the company, with long distances to the small primary school and clinic along poor roads, and the reduction in land as they are squeezed by project planting expansion. Interviewees in Lwanika and Budhala commented on the degradation of conditions at the villages in comparison to 1999, when a report by Makerere University Academics documented their livelihoods (UFRIC 1999).

'In 99 people were better off, could practise fishing, and also cultivate maize on the unplanted land; we also had a smaller population. Evictions have brought about limited land, and all which was taken away was planted so there is almost no land left. It is almost now famine' (Local Councellor (LC), interview with the first author, Lwanika, July 2012).

The Mt Elgon Regional Ecosystem Conservation Programme (MERECP) prospective REDD+ project

Contrastingly, the Mt Elgon Restoration Conservation Project (MERECP) has a very different spatial orientation in its regional approach to the conservation of the Mt Elgon water catchment area, the primary feeder system for Lake Victoria and the Nile itself (LTS International 2011). Though not as yet a certified carbon forestry project, the project has submitted terms of reference for Project Idea Note (PIN), which is the first step in a lengthy REDD+ project formalisation process. Additionally, during a meeting in August 2012 in Kampala with the National REDD secretariat and working group, there was discussion about 'demonstration' REDD activity status for the project.

The MERECP project's aims were initially ambitious in its scope. However, as a donor-funded project there was some dissatisfaction on the part of the donor partner Norway with the project initiators – the International Union for the Conservation of Nature (IUCN) – with the costs of its administration far outweighing the benefits in the ground as they saw it. After its Mid-Term Review, the project was reformulated under the Lake Victoria Basin Commission and it was suggested that the project be scaled down and more targeted, reducing the scope to one sub-county each in three districts on the Ugandan side, and aiming at working closely with community groups in these localities. The objective of the project was to protect the transboundary Mt Elgon National Park, the gazettement of which caused a large degree of social conflict (as discussed in the next section), and as such a core component of the MERECP work was to target this dynamic in the project area, mediating boundary conflicts between the Uganda Wildlife Authority (UWA) and specific community-based organisations.

The project seeks to improve livelihoods for the neighbouring communities and to protect the park through the removal of the need to encroach, or at least provide incentives to avoid doing so. However, it is certainly evident that the community organisations were central to the mediation of the conflict, and that, to a large degree, the livelihoods improvements augmented and expanded by MERECP were already in emergence due to the activities of the groups. Additionally, the incremental improvements in relationships between these groups and the Wildlife Authority were further lubricated by the MERECP funding. Two of the nine groups involved in MERECP in Uganda, in Wanale Subcounty, demonstrate this dynamic, namely the Bushiuyo Dairy Development Group (hereafter called the Bushioyo Dairy Group) and the Budwale Honey Enhancement Group, initiated by the former in 2008. Both began from humble origins as lending groups with meagre funds. In 2007, the Bushioyo Dairy Group, after realising some benefits and accumulating knowledge of proposal processes, applied for a 20% revenue sharing grant from the Wildlife Authority's park gate takings.[2] This grant was to purchase cattle for the group, which they augmented with a small agricultural loan. With the animal manure for short-term crops such as cabbage, potatoes, peas (needed to make a turnover to keep the benefits coming in the short-term before longer-term dairy development) they were able to 'decrease their appetite for encroachment' and improve the relationship with UWA staff in the area. The Budwale honey group followed suit in 2008.

By 2009, MERECP, after seeing the progress made by these two groups, gave 20 million shillings (approx. US$ 20,000) as a community revolving fund (one component of MERECP's activities) to the Bushiuyo Dairy Group for dispersal to community members at low interest rates. Investments from this fund have thus far included 30 dairy animals, cash loans to the group and loans to 15 individuals for agroforestry woodlots who decided that as the money was from the forest they wanted to assist to mitigate the climate. Now in its second loan cycle, the revolving fund has increased from 20 to 24 million Ugandan shillings. The Budwale Honey Group subsequently qualified for alternate MERECP project components. These included a significant contract to do 'enrichment planting' in the Park boundary, which employed locals to reforest degraded areas of the park with native species, and a carbon sequestration scheme to pay selected villagers for sequestering carbon through agroforestry. These activities are not however present in all the sub-counties and nine community group's working relationships, with revolving funds in only three of the nine Ugandan sites. The most visually obvious impact of the project in Wanale sub-county, under the management of both community organisations is the MERECP 'livelihoods plantation'. This is a 20-m wide boundary of eucalyptus that acts as a demarcation to separate the National Park along its (contested) 1993 park boundary and the 'sustainable use zone' which comprised the land under dispute and which must by agreement with MERECP be managed sustainably. This buffer between the communities and the park is impressively visible, and the communities own the trees and share in the revenues of their 'thinning' – 70% to the community fund and 30% to the neighbouring sub-counties. The members show pride in their achievements and their area (the community organisations even renovated the road along Wanale) and attest that they themselves now protect the park from encroachment, and work to sensitise the local community about its benefits.

Historicising 'encroachment' and the contestation of space

Historicising both protected areas and project boundaries provides a fuller picture of the evolution of control over those territories which were only relatively recently taken up and inscribed as the project localities. This provides a view to a very different picture than the project framings themselves allow. It also re-politicises the spatial and social terrain to allow a contextualised examination of the contestations over the spaces of sequestration foreclosed by the *de jure* legitimacy of the projects' claims to state protected land, and the actors and activities these both legitimate and exclude.

A number of the diverse communities in Mt Elgon contest what they consider the customary lands bordering the National Park and specifically a boundary dispute deriving from the 1993 gazettement and renaming of the Park from its prior designation as the Nkokonjuli forest reserve (gazetted in 1964 and governed under the Forestry Department until 2003). According to the Ugandan Wildlife Authority (UWA) this was for its environmental protection, which was necessary due to population pressures and unsustainable land use practices on the area's steep slopes neighbouring it. The disputes stem primarily from the inadequately marked boundaries, which UWA officials themselves attest was problematic (using chain and compass methods). As the communities see it, since 2002/2003 the UWA attempted to 'encroach onto public land' (Namisindwa Land Owners Association members, interview with the first author, July 2012), to which they had customary claims. The communities the first author spoke to in Namisindwa, for instance, explained that their being prevented from performing circumcision rites on the graves of their ancestors (which is seen as necessary in order to fulfil cultural obligations) has resulted in their sons' inability to marry or produce children. This fierce defence of claims to land is recorded in ethnographic research by Suzette Heald (1989) on Mugisu masculinity and the 'anger' of the Mugisu people, and is evidenced with the fierce defence of cultural norms such as circumcision.[3]

This 'encroaching into Public land'/'reassertion of boundary' dichotomy has resulted in long-lasting conflict between UWA and residents within the Mt Elgon region. The area – comprising Manafwa, Bududa, Sironko and Mbale counties – has had ongoing boundary conflicts and protracted resistances for over a decade. These have included a number of deaths of both community members and UWA rangers. These followed evictions in 1993, in which there were no consultations or compensation, and in 2002, when the UWA evicted 550 families from Mt Elgon and destroyed their houses and crops. The situation required the 'intervention' of the President in 2010. His assessment was that the 1993 boundary should be followed, but Mr Adonia Bintoora, the community conservation officer at UWA, asserts that this was instead interpreted by various communities as a directive to go back onto the park.

The complexity and multiplicity of boundary contestations since this time have played out differently over the park; for example, communities were asked to respect the 1983 park boundary in the Sebei region and the 1993 boundary in Bugisu. Two of these multiple contestations are particularly relevant to this analysis. The first is one of three lawsuits over the disputed land. After forming in 2004, the Naamisindwa Land Owners Association filed a suit against UWA and reoccupied disputed territory from the centres in which they had been concentrated after the eviction. The suit resulted in an ongoing court injunction on further activities by either party since 2005 pending a reappraisal by surveyors. However, the community assert that after the injunction this decision was not respected by the UWA until a human rights network officer began to intercede. This resulted in the removal of problematic staff at UWA. The second case concerns the MERECP sites (still admittedly very few) where the project has enabled communities to successfully mediate conflicts and tacitly retain circumscribed access to the disputed territory, reinscribed as a 'sustainable use zone'. This approach, ostensibly under the prospective REDD+ project, contrasts strongly to the engagement that is evident with communities at Bukaleba.

The Tsetse flies in the statue

At Bukaleba Forest Reserve the *de jure* legal context at first appears clear. The only contestation, or 'encroachment' as project implementers and Forestry Authority officials would have it, is due to relatively recent migrations, stemming from the political turmoil over successive regimes and the civil war and the repression and disruption under Amin in the 1970s. This reflexively legitimises the continued territorial integrity of the reserve. However, delving deeper into both the history of the area, and that of the villages predating the project and forest reserve, reveals a more complex process of community disenfranchisement in the face of the colonial state. Whilst making a case for individual community claims to land is beyond the scope of this analysis, it is clear that the legacies of these villages as spaces of inhabitation (some predating the colonial period) have contributed to the current plight of the communities and their *de facto* claims to the meagre, circumscribed livelihoods they sustain under the *de jure* legitimacy of the project. Some of these historical factors predate the colonial period, other more recent ones concern their interaction with external factors such as poll-booth politics[4] and external land pressures leading ultimately to their being trapped 'between a rock and a hard place' (FERN 2000).

Villager's oral histories corroborate the aforementioned 1999 UFRIC report which traces the origins of the Bukaleba Forest reserve to the late 1930s, i.e., during the colonial period (UFRIC 1999). Political and ecological control of the Bukaleba forest along with its inhabitants was cemented after an episode of sleeping sickness in 1939. As a consequence, residents of the whole area were removed under the East African Tsetse Fly Control (EATFC) Acts.[5] As part of this, hunting and eating of game meat were banned. The communities were promised they could return once the situation was under control, but their return was fragmented by large numbers of deaths, departures north to towns such as Soroti, and further control measures over three periods of infestation, culminating in the gazetting of the reserve in 1974 as a Forest Reserve.[6]

With the politically demarcated territory of the reserve formalised through tsetse fly control activities, communities continued to plant areas and 'encroachment' persisted until between 1989–1999, when people were evicted from the forest without compensation. Those who resisted were imprisoned and it is alleged that some were killed in the process. The aforementioned report (UFRIC 1999) mentions starvation and suffering, loss of property and over-crowding as a result of the evictions in villages such as Lwanika, Nakaanga and Walumbe. It highlights the inevitable tensions between the Forest Department personnel and the neighbouring local communities, despite the fact that the communities had tacit settlement and agricultural ties to the land demarcated as forest reserve. These tensions continued after 1997 when a prior planting project by the Norwegian 'Tree Farms' began. Jutta Kill, a member of FERN, wrote in 2003 that the project: 'threatened to evict some 7000 people living on the land to be turned into a carbon offset project. International criticism at the time stopped the project from claiming carbon credits to 'offset' a Norwegian power plant's emissions, but nonetheless, the project continued and trees were planted' (WRM 2003, 1).

Delving even further, the local elder at Lwanika relates a village story that indirectly attributes Colonial agency to the dispossession of the land from communities. The story relates to Bishop Hannington, ordained as the first Bishop of Equatorial Africa in 1885. He arrived by boat, landing at Lwanika – bordering the contemporary Bukaleba – where he was captured; for his part in what was percolating to be a conflict between the Muslim, Catholic and Protestant missionaries for influence in the Buganda state. The then Kabaka ('king') of the Buganda was Mawanga II, a fiery and tenacious character, who saw his influence over his state under threat from foreigners. He ordered the execution of the Bishop, whose last words to the soldiers who killed him were allegedly to inform Mwanga that he had purchased the road to Uganda 'with my [Hannington's] blood' (Reason 1978).

The story follows that years later, in what could be interpreted as a symbolic claim to new colonial territory, a statue of Hannington was erected at the site of his death near Bukaleba. The local population thought that there were minerals hidden in the statue and went to smash it down to retrieve them. What they found however was that as a trick 'the colonists' had instead placed Tsetse flies in the statue that proceeded to visit a plague of revenge upon the Baganda for the killing of the Bishop. Such stories may be conceptualised as 'selectively factual' narrative devices or idioms for local understandings, discussion of, and sometimes action on, or in response to, processes of social transformation. These range from marginalisation to prosperity, and may include changes in structural conditions, attribution of actor agency or other associated agencies. Furthermore, it is also accurate in its attribution of colonial agency to the incidence of tsetse fly outbreak, as the next section will demonstrate; where implicating the colonial state in the ecological control of territory is directly linked to the opportunities for dispossession and enclosure.

Narrative 'friction' and the contemporary spatial state assemblage

With an idea of the complex histories of both sites in place and a politicised appreciation of the orientations of the actors present in the respective landscapes, it is possible to reassess project claims to legitimacy and the spatial implications of implementation. In doing so we move beyond project framings to consider the broader interactions that comprise the spaces of sequestration.

As an initial observation, both projects work to re-enforce the claims of the state, or to be more specific, particular parts of the state apparatus, to territory. Both projects seek to bolster the respective jurisdictions of the NFA and UWA in the Bukaleba Reserve and Mt Elgon National Park, which is an attempt to secure the internal territorialisation of the State, an element of territorialisation often unconsidered in the social sciences (Vandergeest and Peluso 1995). This re-emphasises exclusionary 'fortress conservation' approaches, albeit those which the MERECP project works to mitigate, whilst at the same time asserting at least rhetorical attention to community 'development' and management practices (more on this shortly). The project histories however, which are reinterpreted by the respective project developers, adopt very specific temporal views of the bounded territories they adopt.

In the Mt Elgon context, the renaming of the Park at the time of its conversion from the forest reserve became the symbolic starting point from which current claims by UWA to the park's integrity are drawn, or transferred from the previous park. According to a senior member of the Bushiuyo Dairy Group, there was a realisation of the necessity of a name change to accompany its new dispensation. This was because, as he saw it, parliamentarians appreciated that communities had a historical claim to the area. Successively alienating those claims through both the act of holding the lands 'in trust' for the people by the colonial, and subsequently post-colonial governments, and then through the reinscription as a renamed national park discursively rendered the land available as a national park.

Similarly, the starting point for the Bukaleba project, with its inception in 2006, draws its legitimacy from the park gazettement in 1974. These framings delegitimise prior land claims, such as villagers' symbolic and cultural claims to the land. For instance, the village of Walumbe was settled in 1972, two years before the gazettement, near a giant Mvule tree (Chlorophora Excelsa) that for villagers houses the spirit of Walumbe (meaning Death in Luganda), the brother-in-law of Kintu, the Baganda founding father. It is from this tree that the village name is drawn, and which they attest is a sacred site for Kalichoro believers. Similarly, at Lwanika there is an 'Enkuni Shrine' which superseded a large Mvule tree, that was destroyed by lightning in 1999 (UFRIC 1999), and the sacred initiation/circumcision sites and ancestral graves in Wanale and Nakaalanga in Mt Elgon. These sites, as identifiers of place, form a crucial part of people's relationship to the land, not as a static entity, but as an ongoing social (re)production in the face of fluxes, including the incoming of migrants and the challenges they face with the company.

These framings of community 'place' and project 'space' come into 'friction' (Tsing 2005) with each other. The narrative reworkings of the project are thus necessarily selective and incomplete. The BFC has for instance inadvertently given tacit encouragement to encroachers; through employment of the cheap labour of individuals from the villages within the reserve – which allows the company to claim 'development benefits' for the local communities. In the case of Walumbe village, the company avoided cutting out the Mvule tree in deference to the community and (it can be assumed) the potential conflict it could entail. Additionally perhaps the most subtly pernicious friction is that the BFC and the carbon project owe their presence in the Bukaleba CFR to the 'degraded' state of the forest reserve; a situation pertaining from its 'encroachment'. This is because Bukaleba's official designation under the NFA's 'Forest Functions and Classification' of 2005 (delineating which activities may be undertaken in specific CFRs) as a CFR of 'ecological and biodiversity importance' because of its proximity to the lakeshore. Although the classification allows the 'sustainable use' of a designated 'production zone' in the CFR, this should preclude industrial forestry activities which have only gone forward as a 'pragmatic' course of action in the face of the 'degradation'. Frictions are thus also productive (Tsing 2005), reflexively enabling the discourse of 'degraded' or 'empty' lands that can then be taken up by the project just as they marginalise communities in the process.

Particular actor–state assemblages

Just as new spatial discourses work to de-emphasise or discard prior claims to land, they rescript project 'locality' narratives for particular ends; and for particular assemblages of the state apparatus and other private and non-governmental actors. We say 'particular assemblages' both to emphasise and implicate the central state in its contemporary assemblage as arbiter of claims to territory, and to simultaneously pluralise the conception of 'state' to encapsulate the specific alignments of actors: in the project cases; logging interests and the quasi parastatal NFA, and donor/NGO networks with UWA. This type of assemblage analysis (DeLanda 2011) is gaining traction in multi-scalar analysis, and we suggest it could be of particular importance to the assemblings of 'Green Economy' and the associated costs they might entail such as the carbon forestry projects considered here.

It is easiest to identify the continuity of ecological control by such assemblages through formal legal challenges of the selective, place-based reinterpretations of legitimacy in the project localities. These challenges have included lawsuits in three of the districts (non-MERECP it must be noted) around Mt Elgon, and at least two explicit carbon forestry cases, relating to the New Forests Namwasa Reserve mentioned in the introduction. Both sets of suits assert claims to inter-legality and overlapping sets of laws and constitutional rights, challenging the specific claims to territorial sovereignty of the NFA and the UWA. The Namisindwa Land Association is contesting customary rights, conveyed through long use, to the disputed land between the successive MENP boundaries and title issued under the Amin regime. The appeals under CFR land are bolstered by research by the Ugandan civil society organisation Advocates Coalition for Development and Environment into the conflict between the NFA's management of forest reserves and the rights of those occupying that land which concludes that the Forestry and Tree Planting Act conflicts with other provisions of National Law. As Mugenyi (2005, 26) notes:

> It is legally conceded that uncontested long possession of land ultimately confers legal title to the occupier irrespective of the formal or documentary record of ownership. This common law principle forms the foundation of the bona fide occupants recognised under the Land Act of 1998.

Such provisions of the Land Act are precluded from applying to Protected Area lands however, under the provisions of the National Forestry and Tree Planting Act of 2003. This was promulgated under the Forestry Sector Reform process in the early 2000s that led to the creation of the National Forestry Authority and the National Forestry and Tree Planting Act of 2003. Hybrid actors in assemblage analysis are important for the multiple roles they play, and during the reform process a number of such actors were also working in roles linked to carbon forestry, or for instance in an advisory capacity to the Presidential advisory process on how to deal with encroachment in CFRs during the same period. A letter from one such actor emphasising this fact in the face of contestations cites 'The [Central Forest Reserves] are held in trust for the people of Uganda (those are my words in Section 5 of the Act) – nobody is allowed to "own" land in them, or even to live in them, unless it is part of the forestry purpose that has been licensed' (Letter to Dame Barbara Stocking (Oxfam) by the Chairman of the Uganda Carbon Bureau, October 2011). Whilst such an assemblage and indeed worldview hold '*de jure*' legitimacy, and can be legitimated at the national scale through the perceived need to build 'the forestry industry', social theorists are charged to ask at what costs are these assemblings and their often asymmetrical processes and results achieved? And what alternative land uses, and land histories in this case, are precluded?

A new practice?

The type of spatialising process we are discussing has evident historical precedent, and has been fundamental to the emergence of the modern territorial state in the region. Indeed, Neumann (2004) attributes the British colonial administration's efforts to reorder 'nature', production and human society within its East African territories as a process of state-building, entailing an enclosure of the commons enacted under its 'civilising mission'. The spatial strategy employed to make its far flung citizens 'visible', contained and ordered are one side of the same process as the bounding and containing of 'wild nature' in parks and reserves, creating 'wilderness' spaces, of which MENP and Bukaleba CFR are a derivative. These are not vestigial 'pre-modern' spaces but rather a product of modernisation efforts, and as such a critical component of the 'landscape of modernity' (Neumann 2004). Such perspectives from historical ecology reveal that all environments have long and complex human histories and are anthropogenic to some degree (Erikson 2008).

Drawing this into the context of Bukaleba, such perspectives add explanatory weight to the local narrative of the tsetse flies in the statue that effectively attributes colonial agency to tsetse fly outbreaks and the opportunities for subsequent control and local dispossession that followed them (Ford 1971). In this vein, Reader (1999) and Kjekshus (1977) both locate the disruption of local systems of ecological control (and hence local pre-colonial economies and livelihoods) by colonial authorities to the outbreak of tsetse flies and Trypansomiasis. They assert that the disruption of existing grazing patterns in East Africa directly contributed to the regrowth of the bush lands which harboured explosions in tsetse fly numbers. Reader (1999) attributes the death of 90% of East African cattle between 1900 and 1910 to the spread of colonial systems of ecological control.[7] Finally, the legacy of Idi Amin Dada is a crucial dynamic to reconcile with the spatial understanding of carbon forestry and its selective interaction with the Ugandan state assemblage. His declaration of an economic war to double agricultural production under the Land Reform Decree of 1975 announced that every Ugandan was free to settle in any part of the country. The IUCN has estimated that 25,000 ha were encroached as result of Amin and Obote II (IUCN official, Interview with the first author, August 2012, Kampala) and subsequent mismanagement has exacerbated the problem. Thus, in Mt Elgon, people will invoke the Amin period when justifying their presence through the licences allocated during that time (such as those held by Namisindwa Land Alliance) but which they are retroactively told is illegal, illegitimate or fake by UWA or the NFA in alignment with the post-1990 NRM government. Projects themselves, imbricated in the new assemblage, invoke this period when citing encroachment problems and seek to delegitimise the prior claims of the Amin state to territory which work against their interests as the following assertion indicates: 'Inevitably this resulted in mass encroachment of CFRs, and successive governments have struggled to reverse this action' (Green Resources 2010, 31).

The problem with this inconsistency, and in terms of the frictions at the project level outlined above, is that communities are caught up in 'state schizophrenia'. Whilst Bukaleba residents interviewed by the first author base their claims for compensation on long residence, the NFA and BFC say they cannot formally recognise 'illegal encroachers'. This 'illegality' is not only stopping the allocation of the 500 ha of community managed CFM land, but is also having other pervasive effects. According to the assistant district forest officer, communities in the reserve do not get full government services, or NGO support (apart from donor water pumps in two villages; interview by the first author, July 2012). To live as such is to exist in an interstice, a space of unacknowledged contestation.

Conclusion

We have examined the framings of both projects and explored how they work in different ways to re-emphasise the claims of differentiated assemblages of actor–state networks to control over protected area territories. These assemblages are bolstered by claims to the sequestrations of carbon and the conservation of forestry resources at the project sites, which transnationalise local spaces (Neumann 2004) in an attempt to externalise, or ecologically fix (Harvey 2009) the costs of climate mitigation. In questioning where these costs then fall, this paper considered the project's respective interactions with communities on and within their borders. Community management practices, dovetailing with state actor involvement, engendered through community forestry management (CFM) arrangements have been shown to be non-existent at Bukaleba and limited or only selectively successful at Mt Elgon. The MERECP project, in what Lewis (2009) would call a 'progressive space of neoliberalism' has a more positive alignment in empowering local interests. This is evidenced in their attempts to facilitate access to the disputed territory under the banner of the 'sustainability zone', promoting multifunctional landscapes and direct forest benefits. However, the MERECP example also shows how limited in scope such activities are, which themselves are far from a-political. Benefits here are dependent on limited project funds, good relations with UWA and the ability to conform to a specific eco-governmentality (Agrawal 2005) through organisation into community organisations with sufficient collateral and evidence of suitable financial management capacities. Sections of communities are unable to meet these requirements and thus 'fall through the gaps'.

When looking beyond the myopic lens of community benefits however it is evident that these project entities exhibit an uneasy fit with the legacies of the land in their respective protected area 'localities'. This is made most visible by the resulting contestations at both sites. This can be either through direct conflict and challenge to the protected area's territoriality as at Mt Elgon, or through communities' subtle and variegated strategies of resistance in simply making do as best they can. This is not only reflected at Bukaleba but at other A/R CDM-oriented projects as well (CCS 2012).

These difficulties in the case of Uganda are particularly evident during election time. Klopp (2012) has noted that democracy is one of the major drivers of deforestation in Kenya, and a similar assessment could be made of Uganda where multiple disparate localities or sites of democratic state control 'on the ground' come into friction with vote buying and opportunities for patronage. These dynamics reveal the complexity and imbrication of both the contemporary African state and its attendant tension within carbon forestry, and the neoliberal environmental management constellations (Peluso 2007) they comprise.

As we have seen, the problem framing of the projects offers a simplification of messy dynamics that comprise their project sites. A holistic appreciation of the interrelated complexities within the project areas entails a movement away from simplistic characterisations of 'encroachment' to a broader problem set which points to multifaceted pressures on protected areas which include, but are not limited to

- Declining livelihoods needs exacerbated by declining land availability in Mayuge (an average land holding of only one acre in the district, against the national average of 2.6) and Mt Elgon.
- Population and agricultural pressures, exacerbated in Mayuge district by a voracious sugarcane industry[8] receiving prioritisation within National development plans, and at Mt Elgon due to coffee price fluctuations (Sassen et al. 2013).
- A development-fortress conservation nexus, with limited human and financial resources, within which the active promotion of industrial plantations and logging interests – such as at Bukaleba – precludes a focus on district forestry and community engagement.
- Unclear boundary demarcation, limited UWA/NFA capabilities and disrupted/inconsistent conservation approaches over the years
- Issues compounded by a local politics which both amounts to vote buying and falsely purports to take local needs into account.

Finally, whilst the experiences discussed in the two case study projects cannot be said to be representative of all possible carbon forestry projects, unpacking the evolution of two such projects can reveal the full implications, commonalities and differences of their local variations in all their complexity. Understanding these spatial orientations allows the viewer to traverse the myriad of interactions and processes that projects are enwrapped in and take up as a part of their activities, and make visible the complexities (both in continuity and originality) that link old land and new practice. One aspect that stands out is that failing to acknowledge the contested legacies of the land in project localities and to adequately engage with communities by merely painting them as perpetrators does not reconcile socio-political histories with current realities. Here communities are not dealt with, given their situation. They are continually marginalised.

Notes

1. Critics of carbon forestry assert that market environmentalism underpinning carbon forestry cannot be seen to amount to effective climate mitigation whilst simultaneously attempting to reconcile with large industrial emitters, or as is relevant to New Forests, large industrial plantations (CCS 2012). This is not to say that profit maximisation is the only motivation for carbon forestry. Indeed, as Bakker (2005) has pointed out, there are broad array of goals and a variety of social, cultural and environmental factors in the neoliberalisation of nature.
2. The BDDG was granted 2.7 million shillings (UGX) which they used to purchase 20 cows (which were given to 10 beneficiaries who then passed on the calves produced to other members) to share between its members, and to fertilise with manure their fields which before that date had poor soil fertility.
3. This was evidenced by the xenophobic forced circumcisions of 'outsiders' by gangs of youths in Mbale in 2012 (on the basis that 'outsiders' working there should also likewise be circumcised).
4. Where politicians promise the land in return for votes; notable in the 1980s and more recently in 2006 where a minister famously stated 'trees don't vote, people do'.
5. Bans on hunting and eating of game meat were also imposed at this time.
6. In a VCS document (DNV 2012), the BFC claims the forest was gazetted in 1948, however the UFRIC report and villagers put the date at 1974.

7. This sort of contextualised colonial history has its corollary in the relationship between colonial agriculture and malaria proliferation in Egypt, documented in Timothy Mitchell's book entitled the 'Rule of Experts' (2002).
8. The District Environment officer for Mayuge asserts that district(s) with sugar production tend to be poorer in comparison to other regions of the country (Interview by the first author, July 2012).

Notes on contributors

Adrian Nel is a PhD student in the Department of Geography at the University of Otago, New Zealand. He is currently engaged in writing up his Ph.D. thesis on carbon forestry project emergence and forestry governance in East Africa, having returned from six months of fieldwork in Uganda in 2012. He has contributed a co-authored chapter to a report on CDM activities in Africa (Centre for Civil Society, 2012), a chapter to a book entitled Green Economics: Voices of Africa (Green Economics Institute 2012), and has a forthcoming article in Capitalism, Nature, Socialism on a survey of African Carbon Forestry projects and their orientations.

Douglas Hill is a Senior Lecturer in Development Studies in the Department of Geography, University of Otago, New Zealand. Among other things, he has published extensively on issues related to environmental contestation and rural development at a variety of scales.

References

Agrawal, Arun. 2005. *Environmentality: Technologies of Government and the Making of Subjects.* Durham, NC: Duke University Press.

Agrawal, Anil, and Sunitra Narain. 1991. *Global Warming in an Unequal World.* New Delhi: Centre for Science and Environment.

Alcorn, Janis, Asukile Kajuni, and Bob Winterbottom. 2002. "Assessment of CBNRM Best Practices in Tanzania: EPIQ." A report presented to USAID/Africa Bureau-Office of Sustainable Development.

Anti-Corruption Coalition Uganda. 2010. *Namanve Forest Report: Environmental Crisis Looms as Forests Come under Threat: Cases of Forest Giveaway and Illegal Activity.* Kampala: ACCU.

Bakker, Karen. 2005. "Neoliberalizing Nature? Market Environmentalism in Water Supply in England and Wales." *Annals of the Association of American Geographers* 95 (3): 542–565. doi:10.1111/j.1467-8306.2005.00474.x.

Brenner, Niel. Jamie Peck, and Nik Theodore. 2010. "After Neoliberalization?" *Globalizations* 7 (3): 327–345. doi:10.1080/14747731003669669.

Brockington, Dan, and James J. Igoe. 2007. "Neoliberal Conservation: A Brief Introduction." *Conservation and Society* 5 (4): 432–449.

Bumpus, A. G., and D. M. Liverman 2008. "Accumulation by Decarbonization and the Governance of Carbon Offsets." *Economic Geography*, 84 (2): 127–155. doi:10.1111/j.1944-8287.2008.tb00401.x.

CCS (Centre for Civil Society). 2012. *The CDM in Africa Cannot Deliver the Money.* University of KwaZulu Natal Centre for Civil Society (SA) and Dartmouth College Climate Justice Research Project (USA).

Castree, Noel. 2003. "Commodifying What Nature?" *Progress in Human Geography* 27 (3): 273–297. doi:10.1191/0309132503ph428oa.

Castree, Noel. 2008. "Neoliberalising Nature: The Logics of Deregulation and Reregulation." *Environment and Planning A* 40 (1): 131–152. doi:10.1068/a3999.

DeLanda, Manuel. 2011. *Assemblage Theory, Society, and Deleuze.* Saas-Fee: Public open lecture. European Graduate School EGS Media and Communication Studies department program.

DNV. 2012. "DNV Climate Change Services AS (Norway) Bukaleba Forest Proejct in Uganda." VCS Verification Report No 2011-9524 Revision No. 02.

Erikson, Clark. 2008. "Amazonia: The Historical Ecology of a Domesticated Landscape." In *The Handbook of South American Archaeology III*, edited by Helaine Silverman and William H. Isbell, 157–183. New York: Springer.

FERN. 2000. "Tree Trouble. A Compilation of Testimonies on the Negative Impact of Large-scale Monoculture Tree Plantations." Report prepared for the sixth Conference of the Parties of the Framework Convention on Climate Change. Friends of the Earth International in cooperation with the World Rainforest Movement and FERN.

Ferguson, James. 2004. *The Anti-politics Machine - Developemnt, Depoliticisation and Beurocratic Power in Lesotho*. Minneapolis: Minnesota University Press.

Ford, John. 1971. *The Role of the Trypanosomiases in African Ecology. A Study of the Tsetse Fly Problem*. Oxford: Clarendon Press.

Goldman, Michael. 2004. "Eco-governmentality and Other Transnational Practices of a 'Green' World Bank." In *Liberation Ecologies: Environment, Development, Social Movements*, edited by R. Peet and M. Watts, 166–192. Oxford: Taylor and Francis.

Green Resources. 2010. *Bukaleba Forest Project Uganda: Project Idea Note*. Oslo: Green Resources.

Harvey, David. 2009. "The 'New' Imperialism: Accumulation by Dispossession." *Socialist Register* 40 (40): 63–87. https://www.mediatropes.com/index.php/srv/article/view/5811#. UZTFzKU7WAg

Heald, Suzette. 1989. *Controlling Anger: The Sociology of Gisu Violence*. Manchester: Manchester University Press.

Heynen, Nik. 2007. *Neoliberal Environments: False Promises and Unnatural Consequences*. New York: Routledge.

Kjekshus, Helge. 1977. *Ecology Control and Economic Development in East African History: The Case of Tanganyika 1850–1950*. London: Heinemann.

Klopp, Jaqueline. 2012. "Deforestation and Democratization: Patronage, Politics and Forests in Kenya." *Journal of Eastern African Studies* 6 (2): 351–370. doi:10.1080/17531055.2012. 669577.

Lewis, Nick. 2009. "Progressive Spaces of Neoliberalism?" *Asia Pacific Viewpoint* 50 (2): 113–119. doi:10.1111/j.1467-8373.2009.01387.x.

LTS International. 2011. *Mount Elgon Regional Ecosystem Conservation Programme: End Review Report*. Penicuik: LTS.

McCarthy, James, and Scott Prudham. 2004. "Neoliberal Nature and the Nature of Neoliberalism." *Geoforum* 35 (3): 275–284. doi:10.1016/j.geoforum.2003.07.003.

Mitchell, Timothy. 2002. *Rule of Experts: Egypt, Techno-Politics, Modernity*. Berkeley, CA: The University of California Press.

Mugenyi, Onesumus. 2005. *Balancing Nature Conservation and Livelihoods: A Legal Analysis of Forestry Evictions by the National Forestry Authority*. ACODE (Advocates coalitions for development and Environment) Policy Briefing Paper no. 13, Kampala.

Murphree, Marshall W. 2001. *Community Based Conservation: Old Ways, New Myths and Enduring Challenges*. Tanzania Wildlife Discussion Papers No. 29. Dar es Salaam: GTZ.

Neumann, Roderick P. 2004. "Nature-state-territory: Toward a Critical Theorization of Conservation Enclosures." In *Liberation Ecologies: Environment, Development, Social Movements*. 2nd ed., edited by R. Peet and M. Watts, 195–217. New York: Routledge.

NFA. 2009. *National Forestry Authority Business Plan 2012–2014*. Kampala: National Forestry Authority.

Oels, Angela. 2005. "Rendering Climate Change Governable: From Biopower to Advanced Liberal Government?" *Journal of Environmental Policy & Planning* 7 (3): 185–207. doi:10.1080/15239080500339661.

Peluso, Nancy L. 2007. *Enclosure and Privatization of Neoliberal Environments: Some Comments*. London: Routledge.

Reader, John. 1999. *Africa: The Biography of the Continent*. Pennsylvania State University: Vantage Press.

Reason, Joyce. 1978. *Bishop Jim: The Story of James Hannington*. London: James Clark Company.

Ribot, Jesse. 2002. *Democratic Decentralization of National Resources: Institutionalizing Popular Participation*. Washington, DC: World Resources Institute.

Sassen, Saskia. 2005. "When National Territory is Home to the Global: Old Borders to Novel Borderings." *New Political Economy* 10 (4): 523–541. doi:10.1080/13563460500344476.

Sassen, Mareike, Douglas Sheil, Ken Giller, and Cajo ter Braak. 2013. "Complex Contexts and Dynamic Drivers: Understanding Four Decades of Forest Loss and Recovery in an East African Protected Area." *Biological Conservation* 159: 257–268. doi:10.1016/j.biocon.2012.12.003.

Tipa, Gail, and Richard Welch. 2006. "Comanagement of Natural Resources: Issues of Definition from an Indigenous Community Perspective." *The Journal of Applied Behavioral Science* 42 (3): 373–391. doi:10.1177/0021886306287738.

Tsing, Anna. 2005. *Friction: An Ethnography of Global Conneciton*. New Jersey, NJ: Princeton University Press.

UFRIC. 1999. "A Site Report Prepared for Presentation to the Local People of Lwanika, Budhala and Forest Department Office, Iganga District," Uganda Uganda Forestry Resources and Institutions Centre (UFRIC) Research Note Number 3.

URN (Uganda Radio Network). 2011. "Mayuge Residents Clear Bukaleba Forest Reserve." http://ugandaradionetwork.com/a/story.php?s=31411

Vandergeest, Peter, and Nancy Peluso. 1995. "Territorialization and State Power in Thailand." *Theory and Society* 24 (3): 385–426. doi:10.1007/BF00993352.

WRM. 2003. *World Rainforest Movement Bulletin No 74, September 2003*. Monteveideo: WRM.

Conditioned by neoliberalism: a reassessment of land claim resolutions in the Kruger National Park

Maano Ramutsindela and Medupi Shabangu

Department of Environmental and Geographical Science, University of Cape Town, Cape Town, South Africa

The focus of this paper is on land reform in protected areas where two main national priorities for biodiversity conservation and land restitution have clashed, and subsequently led to a model of settling land claims in protected areas. This model, initiated by the resolution to the Makuleke land claim in the Kruger National Park in 1998, was celebrated nationally and internationally as one of the best options for resolving environmental injustices in the country and as an avenue for developing a mutually beneficial pact between people and parks. Our principal objectives are to account for why and how the Makuleke land claim was settled and hailed as an appropriate model, and why such a model – deemed useful – has been abandoned by the government and conservation lobby groups and agencies. We draw on material from earlier research on this case and recent fieldwork data to argue that the Makuleke model has been abandoned partly because the conditions under which it was developed have changed and also because of the neoliberalisation of protected areas in the country. We conclude that the process of neoliberalising nature goes beyond creating conditions for capital penetration in nature to influence land and agrarian relations.

Introduction: neoliberalising protected nature

Neoliberalism has, since the late 1970s, been increasingly used to give shape and direction to political and economic thought, and to create appropriate management structures and regimes conducive for capital accumulation. Harvey (2005, 2) has succinctly summarised neoliberalism as 'a theory of political economic practices that proposes that human well-being can best be advanced by liberating individual entrepreneurial freedoms and skills within an institutional framework characterised by strong private property rights, free markets and free trade'. A growing body of literature has demonstrated that neoliberalism cannot appropriately be understood by adopting totalizing or monolithic narratives, mainly because the process is varied, and also does not engender identical outcomes (Heynen et al. 2007). These trends can be observed in the domain of the environment where neoliberalism and the environment have been linked historically. The environment has also become a site at, and an avenue through, which capitalism expands and also deepens. Thus, rather than threatening the prosperity of capitalism, environmental impediments have

109

become 'the new source of new forms of accumulation' (Brockington and Duffy 2010, 472). Our focus in this paper is on the alliance between capitalism and conservation that is characterised by the framing of environmental problems in ways that often privilege market-based solutions while at the same time opening up avenues for conservation-based enterprises. These characteristics have become clearer in the Kruger National Park (KNP) where the state and conservation lobby groups have, under different political conditions, consistently considered the locals a threat to conservation. As elsewhere in developing countries (see Brockington and Igoe 2006), locals in South Africa were forcibly removed to give way to the creation of the KNP (Carruthers 1995). In post-apartheid South Africa, land claims by locals evoked the same notions of threat, as we shall see below.

The abandonment of the Makuleke model with respect to other land claims in the KNP and other protected areas in the country begs the question of the conditions under which the model was developed; the assumption underlying the agreement central to the model and the reasons for not replicating what was celebrated as an appropriate solution to land claims in South Africa's protected areas. We draw on earlier research and new fieldwork data on land claims in the KNP to engage these questions. We argue that the reasons for the failure to replicate the Makuleke model go beyond the land claim *per se*. They reveal that conservation-based enterprises are not only contingent on changes in property ownership, but can also be pursued by strategically and momentarily dispensing and withdrawing those rights under the logics of conservation and capitalism. To substantiate our argument, we begin the paper by sharing insights into neoliberalism and conservation before we recount conditions under which the first land claim in South Africa's protected areas was resolved. Thereafter, we engage the meanings and interpretations of the model with an eye on showing how and why those interpretations have shifted over time. In the concluding part of the paper, we highlight the implications of the findings of the paper on literature on capitalism and conservation.

On the intersection between capitalism and conservation

The relationship between conservation and capitalism is not new but has, in recent years, deepened and also undergone significant changes as a result of how environmental problems and solutions are understood or framed, and because of changing market conditions. Initially, capitalism and conservation were treated as antagonistic ideas and practices; hence, the relationships between the two were strained. Nature was to be protected from capitalism, especially from the juggernaut of industrial capitalism that left destruction of nature in its wake. It was this opposition to industrial capitalism that spurred the civil movements towards the creation of national parks (Nash 2001). What was initially a dialectical relationship has turned into a comfortable and growing alliance between capitalism and nature conservation. Drawing on the work of David Harvey, Castree (2008) ascribed this alliance to a strong belief in the private sector's ability to 'fix' environmental problems. These 'environmental fixes' manifest in the promotion of the free market as a vehicle for ecological stewardship (free market environmentalism): opening up protected aspects of nature to the market; intensifying neoliberal measures by extending the rights of the private sector to include the right to exploit nature and by off-loading the state's responsibility to manage the contradictions of capitalism

(Castree 2008). Thus, free market environmentalism presents capitalism as the key to our ecological future that will in turn help end our current financial crisis (Igoe, Neves, and Brockington 2010). Nature can be saved by selling it (McAfee 1999): using the revenue from nature to finance the protection and management of natural resources.

Others see the alliance between capitalism and nature conservation as mediated through consumption (Igoe, Neves, and Brockington 2010; MacDonald 2010). Carrier (2010, 672) has argued that 'ethical consumption marks a conjunction of capitalism and conservation, for it identifies people's market transactions, and market mechanisms generally, as the effective way to bring about protection of the environment'. Such consumption, ethical or not, is dependent on processes of deregulation and reregulation that not only open up access, ownership and management of resources such as wildlife to private actors, but that also enable the commodification of nature. Commodification entails the creation of an economic good through the application of mechanisms to appropriate and standardise a class of goods or services, enabling them to be sold at a price determined through market exchange (Bakker 2005; Büscher et al. 2012).

We would not be in error to suggest that conservation NGOs act as glue between capitalism and conservation; hence, Brockington and Scholfield (2010) have described these NGOs as constituting *a conservationist mode of production* – they help perpetuate the demand for the consumption of wildlife services. The links between capitalism and conservation NGOs can be traced back to the late 19th and early 20th centuries when the elite created environmental non-governmental organisations to press for nature conservation policy reforms that would protect hunting by the elite (Bryant 2009; Ramutsindela, Spierenburg, and Wels 2011). These NGOs carried out the task of creating nature reserves in order to preserve the wild for enjoyment by the elite despite the historical fact that wildlife numbers had plummeted as a result of the lifestyle of the same elite. Accordingly, environmental NGOs are seen as both a product of elite environmentalism and an agency for eco-imperialism (Bryant 2009).

Of immediate relevance to the theme of this paper is the link between capitalism and conservation through commercialisation. As we shall discuss below, commercialisation has a bearing on the resolution of the Makuleke land claim and the response to other land claims in the KNP. Bakker (2005) provides a succinct explanation of commercialisation in the environmental domain when she notes that the notion of commercialisation entails changes in resource management practices which introduce commercial principles, methods and objectives. Such a commercialisation process heavily relies on the principle of efficiency that is used as a discursive trope for reconciling the interests of market enthusiasts and the goals of environmental conservation. The logic here is that natural resources (including national parks) in the hands of the state should be managed efficiently and that the market, rather than the state, has the ability and skills necessary for efficiency. Thus, the principle of efficiency opens up state functions to the private sector.

We referred briefly to the links between capitalism and conservation above to highlight that business interests cannot be discounted from conservation efforts, especially when solutions to environmental problems are couched in market-related terms. Such environmental problems take different forms, including the threat that locals are seen to pose to conservation efforts. In the case reported in this paper, the threat to the conservation status came in the form of land claims against a national park that stands as a beacon of conservation efforts in South Africa and beyond. The solution to this apparent threat has commercial undertones.

Struggles for land rights in the Kruger National Park: how the first battle was won

Reconciling land reform and biodiversity conservation is a complex undertaking, especially in developing countries where deep socio-economic problems and imprints of historical injustices exist alongside some of the richest biodiversity endowments in the world. It is also hard to realise such reconciliation when there is lack of understanding between 'sectors dealing with biodiversity conservation and those dealing with human and land rights' (Kepe, Wynberg, and Ellis 2005, 3). Such lack of understanding leads to mistrust which, in turn, prevents the development of innovative strategies that could balance the goals of land reform and biodiversity conservation. In the South African context, land reform and biodiversity conservation are high on the government's priorities and also enjoy constitutional support. All land claims in South Africa are guided by the constitution of the country, and are given practical legislative expression by the Restitution of Land Rights Act 22 of 1994 that seeks to simultaneously pursue the goals of redress and reconciliation under market conditions. This Act allows for and has been used by victims of racially motivated removals to claim land in urban areas, white farms, rural areas and game reserves and national parks. The varied outcomes of these claims range from land restoration to cash payments. With regard to the land claims in national parks, when the Makuleke lodged the claim in December 1995 that was gazetted in August 1996 (South African Government 1996), government policy was vague on how land claims of this nature were to be resolved. Instead, the policy hinted that the settlement of land claims in national parks would 'combine the objectives of restitution with the conservation and sustainable use of biodiversity' (South African Government 1997, 34).

The Makuleke land claim was to test this pronouncement at the time when the political mood for land reform was very strong and also conducive for a fierce contestation over land under claim. Central to the contestation was the Makuleke's quest for the restoration of full land rights to the community. The community's intention was to resettle at old Makuleke (Pafuri), which meant re-establishing a settlement inside KNP (Mavis Hatlane, interview, 9 February 2012). Pitted against a powerful and globally networked conservation lobby, the Makuleke needed a strong backing to pursue their land claim in the KNP. Such a backing came in the form of the Friends of Makuleke consisting of 'highly qualified and articulate white South African professionals in community development and planning ... [whose] arguments matched the esoteric language of the officials of SANParks' (Magome and Murombedzi 2003, 121). Formally constituted in 1997, a year after the land claim was gazetted, this group provided the Makuleke with technical support throughout the land claim (Spierenburg, Steenkamp, and Wels 2006), while Moray Harthon of the Legal Resource Centre in Johannesburg coordinated the legal backing of the claim. In addition to the land claim proper, Harthon argued for the restoration of mineral rights to the Makuleke or payment for compensation for those rights (Harthon 1998) at the time when there was lack of attention to mineral rights in the country's land reform policy. The provincial Land Claim Commission was sympathetic to the land claim. The former Land Claim Commissioner in the province, Mashile Mokono, commented that, 'for us restitution to Makuleke represented a political moral high ground than the depth of economic logic put forth by conservation lobbyists. Issues of equity, transformation including black economic empowerment were becoming very real to us' (Interview, 12 February 2012).

This kind of social activism, together with heightened political stakes in land reform in the country at the time, gave the Makuleke land claim, in our view, unparalleled support for land claims in protected areas in the country. Such a level of support was also warranted because, as we shall see below, the land claim (if successful) would make a huge dent in the KNP and what the park represents in political and ecological terms (see Carruthers 1995 for details) and would also disrupt the long-term goals for creating transfrontier conservation areas (TFCAs) that were wrapped in secrecy at the time (Ramutsindela 2007). The land claim also required strong support because it elicited a strong opposition, the point we turn to below.

The demand for land restoration by the Makuleke was vehemently opposed by a strange alliance consisting of the South African Department of Land Affairs (DLA), the Wildlife and Environment Society of South Africa (WESSA) and South African National Parks (SANParks), and was joined by international conservation lobby groups. This alliance existed by default as each opposing group was motivated by different objectives that coalesced on the same piece of land under the claim. Ironically, the Department of Land Affairs (DLA) that was driving the land reform programme was opposed to the restoration of full land rights to the Makuleke. In representing the DLA, Mouton argued that the Makuleke did not have formal title or allocation to land but conceded that since they used land in the past, they should be considered for land redistribution or development programmes (Ramutsindela 2002a). Notwithstanding divergent and sometimes conflicting views on the direction of land reform within the DLA, our point is that the government was unsure about how to resolve land claims in protected areas, and government departments were, as in the apartheid era (see Carruthers 1995), in conflict with one another. For example, the Department of Minerals and Energy (DME) had issued prospecting rights for alluvial diamonds in and around the park, an area Ferrar, Wynne, and Johnstone (1997, 16) call the crown jewel of the Kruger National Park. For its part, the Department of Environmental Affairs and Tourism was opposed to mining, considering it a threat to the conservation status of the KNP.

Conservation groups, led by WESSA and SANParks, were at the forefront of the campaign against mining in the area. It should be noted that when the South African Defense Force that was in charge of Madimbo Corridor bordering Zimbabwe lifted the ban on mineral prospecting in the area at the dawn of democracy in 1994, the Australian-based Madimbo Diamond Corporation obtained a prospecting permit in 1995 to the chagrin of environmentalists who saw prospecting as opening up an ecologically sensitive area for destruction. Against this backdrop, WESSA and SANParks saw the Makuleke land claim as another threat to the conservation status of the KNP and opposed the land claim outright. However, this threat by the land claim was later to become an advantage: The victory by the land claimants could be used to put a brake on mining prospecting and also give credibility to the proposals for the Great Limpopo Transfrontier Park (GLTP) on the Mozambique-South

Africa-Zimbabwe border. Thus, disputes over mining and mineral rights and conservation goals became inseparable, and these two converged over the Makuleke land claim. Accordingly, the land claim facilitator, Bosch, concluded that the land claim could potentially be derailed by disputes over mineral rights (see Ramutsindela 2002a). It should be emphasised that at the time, land reform policy was silent about mineral and other rights to natural resources other than land. Under these conditions, interested parties had first to agree that there would be no prospecting or mining in the area before the land claim could be settled. Arguably, conservation groups won the battle against mining by using the very land claim they were initially opposed to. Their second battle was to protect the conservation status of the KNP against the land claim.

As we hinted above, conservation groups saw the Makuleke land claim as a threat to two interrelated conservation goals, namely, maintaining the conservation status of the KNP and pursuing the plan for TFCAs. The argument against the land claim on conservation grounds reflects broader conceptions of the national park and the place of indigenous and other local citizens in national park management systems. It is generally believed that national parks should show a clear separation between humans and nature. The intention by the Makuleke to re-establish their settlement in the KNP through a land claim went against this belief, which is fundamental to the practice of conservation. Beyond securing the conservation status of the KNP, the conservation lobby saw the land claim as a threat to the bigger plan: the establishment of the GLTP. The plan for creating the GLTP predates the land claim and reflects a much wider political history of conservation in southern Africa, and also general trends in conservation thinking (Ramutsindela 2007).

Concerns with ecosystems and the important roles they play in the biophysical environment and society have given impetus to strategies to maintain, protect or even recreate ecological systems. In conservation circles, these strategies include the creation of cross-border conservation areas such as the GLTP that links together the KNP, Limpopo National Park in Mozambique and Gonarezhou National Park in Zimbabwe into a contiguous conservation area, jointly managed by the three countries. Conservation projects of this nature are not new in the region – they were pursued by colonial governments for reasons that have been adequately discussed elsewhere (Mavhunga and Spierenburg 2009). They are also found in other parts of the world. Of relevance to the discussion in this paper is the relationship between the GLTP and the land claim.

The plans to create cross-border conservation areas such as the GLTP in post-apartheid southern Africa were drawn shortly before liberation in South Africa but were never made public, in part because they were politically sensitive (see Ramutsindela 2007). They could be read as an Afrikaner political exit strategy that ensured their control of some part of the land in the region, as evident in the analysis by Stephen Ellis (1994). The GLTP was from the onset the main target for these conservation projects because of the iconic status of the KNP that constitutes the cog of the GLTP. The Makuleke land claim was seen as a threat to the envisaged GLTP in that it could take away land that had already been mapped as part of the GLTP or use that land in contradiction with large-scale conservation plans and the whole purist idea of protected areas.

It follows that both mining and conservation shaped the resolution of the land claim under discussion as shown by the agreement that was signed by nine parties[1] on 30 May 1998 and made an Order of the Court on 15 December 1998. These two land uses are central to the settlement of the claim, with mining issues appearing at the top of the list, as evident in Clause 11 of the Agreement which reads as follows:

> no mining and/or prospecting activities (as defined in the Minerals Act), may take place in, on or under the land…no part of the land may be used for residential purposes insofar as such purposes would conflict with the land being maintained and utilised as an area of conservation and associated commercial activities; no part of the land may be used for agricultural purposes; the land is to be utilised and maintained solely for the purpose of conservation, and associated commercial activities[2]; and no development of whatsoever nature may be made on the land prior to an environmental impact assessment as may be required by law being undertaken and the approval of the competent authority in terms of such law being obtained in respect of such development. (Main Agreement Relating to the Makuleke Land Claim 1998, 6)

As we argued above, the discourse of conservation advanced against the land claim was not limited to protecting the status of the KNP but was also linked to plans for the creation of TFCAs. It is for this reason that Clause 17 of the Agreement bound interested parties to the open ecological systems with neighbouring conservation areas. The idea of an open ecological system in the Agreement was specifically directed at the Makuleke Region of the KNP, portions of the Matshakatini Nature Reserve and the Makuya Park. The language used to argue for such an ecological system was similar to that of TFCAs in general, namely, the free movement of game in a larger area transcending boundaries and the promotion of ecotourism. These goals are clear from Article 4(c) and (d) of the Treaty of the Great Limpopo Transfrontier Park (2002) in which the emphasis is on enhancing ecosystem integrity and natural ecological processes and the promotion of ecotourism. We are mindful that TFCAs have more than two goals (see Ramutsindela 2007), but our focus here is on those that are central to the resolution of the land claim under discussion for analytical clarity.

Conservationists celebrated the way in which the Makuleke land claim was resolved, because the claim neither compromised the KNP nor obstructed the creation of the GLTP. The use of the Makuleke Region of the KNP for ecotourism might appear to be a sober approach to local economic development in impoverished rural communities such as Makuleke, but it also raises profound questions about why such an approach is not used for communities in the same situation as that of Makuleke. More significantly, the approach brings to light questions of the overall business strategy of the KNP. We turn to these questions in the discussion below.

Kruger, land claims and the neoliberal moment

The manner in which the Makuleke land claim was resolved has excited people and institutions of different kinds. These include the Consultancy Africa Intelligence (CAI) that was formed in 2006 to provide its clients with research and intelligence material 'to better understand and respond to African trends, developments, progress and concerns' on issues including conflict and terrorism patterns and threats in Africa, financial and economic risk, foreign relations, human rights and gender-related issues, market penetration, mining and industry, political risk and impacts on foreign investment, public health, etc. The CAI profiled the resolution of the Makuleke land claim to demonstrate how the South African government is managing a potentially volatile land question and also how solutions to the country's teething problems offer economic opportunities (Consultancy Africa Intelligence 2011). To restate our argument, the Makuleke model reflects a high level of support for conservation in the face of demands for social justice through land reform, and also manifest a neoliberal moment in South Africa's protected areas. It should be noted that the settlement of the Makuleke land claim was made two years before the roll-out of the business model in KNP in parallel with commercialisation in provincial game and nature reserves.

Commercialisation in conservation areas in South Africa should be understood in the context of existing governance structures. The statutory mandate for conservation and the management of natural resources is divided into the three spheres of government, namely, national, provincial and local, in line with the overall organisational structure of the state. This means that the business model in conservation in state land is not the same, and that conservation areas under the three spheres of government are not managed together under one comprehensive business strategy for the country as a whole. For example, provincial parks boards whose mandate is to manage provincial nature reserves are, according to Khorommbi Matibe of the Limpopo Tourism and Parks Board (Pers. Comm., 17 March 2011), commercially unviable; hence, they use a strategy to attract private sector capital investment. In 1998 Fish Mahlalela, then Mpumalanga MEC for Environmental Affairs and Tourism, stated that, 'there must be commercial development on our reserves to generate money to fund provincial conservation' (Mpumalanga Parks Board 1998). King (2009) argued that the shifting priorities of government resulted in an emergent commercialisation discourse within the Mpumalanga Parks Board which framed conservation in economic terms that would have meaningful benefit for communities and organisations managing the projects. The Dolphin Agreement that was negotiated between the Dolphin Group and Mpumalanga Parks Board provided enormous potential for a R1 billion investment in provincial reserves. This deal was announced on 27 November 1996 and licensed the foreign group to utilise key attractions within the province, including the Blyde River Canyon, Bourke's Luck Potholes and Pilgrim's Rest. According to news reports, the agreement enabled the Dolphin Group to 'exercise their commercial rights, and maximise development opportunities and financial returns' (Arenstein 1997). The then Premier of the province, Matthews Phosa, hailed the deal as an 'African dream, a unique partnership between the private sector, communities and conservation' (In Gray 1998).

At the national level, the approval of the establishment of a task team by the South African Cabinet in 1997 to explore how Public Private Partnerships (PPPs) could improve infrastructure and service delivery *efficiency* and ensure efficient use of under-utilised state assets marked the accelerated shift to neoliberalism as a central pillar of macro-economic policy. In September 1998, the same year in which the Makuleke Agreement was signed, the Department of Environmental Affairs and Tourism argued that nature conservation and tourism, which are key economic growth sectors, must also pursue such policy shifts. The former Chief Executive Officer of SANPAks, Mavuso Msimang, gave the clearest reason for the commercialisation of South Africa's national parks when he said that, 'the government has a huge responsibility to provide health care, housing, and clean water to historically disadvantaged communities, and so the SANParks is not a huge priority' (cited in

Ramutsindela 2002b, 86). Simply put, SANParks should be prepared for less dependence on state funding (Varghese 2008). At face value, this reasoning makes sense and also chimes with the general view held in conservation circles that African governments have both lack of funds for, and interest in, nature conservation. The argument that the state is unable to fund national parks, because it has huge social responsibilities does not stand the test when held up against the privatisation of the very services the state is supposed to provide (see Bond 2000 on the privatisation of basic services and the consequent urban crisis). The other reason for commercialising national parks in the country is that there are certain services that can best be provided by the private sector. This is clear in the concession contracts that SANParks issued to the private sector where it is stated that, 'the concessionaire has expertise in the provision of accommodation and related services for visitors to the national parks and facilities in connection therewith' (SANParks 2001, 1). The emphasis on the expertise of the private sector clearly demonstrates the principle of efficiency we discussed above.

It follows that the rationales for adopting the Commercialisation Strategy in 2000 was that competing social needs put a strain on the public revenue and nature conservation should leverage private sector investment through tourism development to partly fund its existence. The strategy led to 11 concessions sites, seven of which are in the KNP[3], two in Addo Elephant National Park and two in Table Mountain National Park. All were awarded to private operators. The strategic value of the Commercialisation Strategy 2000 ensured that SANParks captures a greater market segment, increases economic activity and acquires new five-star market facilities built by the private investors on a Built Operate and Transfer (BOT) system. This system means that property built on leased land in national parks would be transferred to the national parks at the expiry of the lease agreement. The net income yielded by this strategy in the first six years is shown in Table 1 below.

To return to our argument, the settlement of the Makuleke land claim cannot be fully understood outside the broad commercialisation strategy that was to unfold in South Africa's national parks. It is not a coincident that the first concession to be granted by the Makuleke community to a private operator was in 2000, the same year in which commercialisation was adopted as a strategy for the operation of South Africa's national parks. The first concession in the Makuleke region of the KNP, Outpost Lodge, brought in R14 million investment (Outpost Concession Deputy Manager, interview, 21 February 2012). Wilderness Safaris' Pafuri Camp

Table 1. Income from commercialisation to SANParks, 2000–2006.

Product	Total income	US$ equivalent (R7.50 = US$1)
Concessions	R54,954,278	$7,327,237
Others	R30,299,941	$4,039,992
Restaurant and retail facilities rental	R37,002,267	$4,933,636
Total	R122,256,486	$16,300,864
Commercial expenses	−R15,254,346	$2,033,913
Net income commercialisation	R107,002,140	$14,266,952

Note: Figures as at 31 March 2007.
Source: Adapted from SANParks (2007).

became the second concession for the Makuleke community and started operating in 2005. This was as a result of a contractual agreement between the Makuleke Communal Property Association (CPA) and the Wilderness Safaris who was the successful bidder for the Pafuri concession. A public-private partnership investment model was developed together with Wilderness Safaris within an area of land that has been and still is the ancestral lands of the Makuleke people. According to Clause 7, Paragraph 3 of the signed agreement that was made an Order of Court, a R25 million infrastructure investment was made by Wilderness Safaris to build the tented camp facility in the northern bank of Luvuvhu River in the Makuleke Region of KNP (Landi Burns, interview, 21 February 2012). The camp is divided into two areas, Pafuri East with 7 tents and Pafuri West with 13 tents. The camp employs 44 members of the Makuleke community and 6 employees from the Wilderness Safaris who are mostly into management and finance. The two concessions articulate a nationally driven system of Build Operate and Transfer (BOT). According to this system, a total of R39 million tourism investments in the Makuleke Contractual Park will revert back to the ownership of the community. This will take place after the lapse of the contractual agreement for the two concessions. It will be upon the Makuleke CPA to decide to run the facilities themselves or open the bidding for private operators (Lamson Maluleke, interview, 7 March 2011).

Tables 2 and 3 show that the Makuleke derived some income from the two concessions. Caution should be taken when interpreting income from concessions as concessionaires might declare low income and losses in order to avoid paying out large dividends. Our interpretation of the performance of concessions in this paper is based on the information that was made available to us. According to such information, the income from the Wilderness Safaris in 2012 was well below projections, and the CPA was of the view that the Pafuri Camp concession was not providing any reasonable returns. This perception is based on comparing Wilderness Safaris' initial projected forecast of R3.1 million as 9% turnover in the fourth year of operation (Koch 2003), with the actual amount of R540,451 that materialised as income paid to the Makuleke CPA as lease fees for 2009 (see Table 2).The Outpost and Wilderness Safaris pride themselves as catering for an exclusive international tourism clientele. The concessionaires believe that the economic recession has significantly impacted on the target market for Wilderness Safaris and The Outpost. With commercial trophy hunting pushed to the sidelines, the only sources of income for the Makuleke CPA are the two concessions. The floods that hit the KNP in February 2013 have worsened the economic prospects of these concessions. They

Table 2. Income (in Rands) from Wilderness Safari to the Makuleke CPA, 2005–2009.

	2005	2006	2007	2008	2009	Total
Wilderness Safaris – projected lease fees to Makuleke CPA (2003)	–	1,523,241	2,100,150	2,689,748	3,166,555	9,479,694
Wilderness Safaris – actual lease to Makuleke CPA	–	230,633	496,836	1,077,532	540,451	2,345,452

Source: Unpublished Makuleke data.

Table 3. Income (in Rands) from Outpost Lodge to the Makuleke CPA, 2005–2009.

	2005	2006	2007	2008	2009	Total
The Outpost – actual lease fees paid to Makuleke CPA	91,007	108,443	210,472	241,458	151,574	802,954
Eco-training – lease fees to Makuleke CPA		21,890	160,655	172,270	180,000	534,815
Total amounts paid by the three concessions	91,007	130,333	371,127	413,728	331,574	1,337,769

Source: Unpublished Makuleke data.

have, in the words of Lamson Maluleke, 'destroyed the cash cow' of the Makuleke community.[4]

The Makuleke model raised questions from the start, but these were ignored till more land claims in the KNP and other national parks came to the surface. The feasibility study that was carried out in the Makuleke region of the KNP had indicated that the area was marginal to ecotourism (Magome and Murombedzi 2003), suggesting the high possibilities for the commercial enterprise in that region to face obstacles long before tourism was affected by the current global economic crisis. We note the view of the Concession Manager at Wilderness Safari, Landi Burns, that tourism is a long-term investment and the realisation of substantial benefits in the case of Makuleke, given the size, dynamics of the tourism market and operations at the Pafuri Camp, will take two to three generations to make an impact at the community level (Interview, 21 February 2012).

Critics, including high-ranking SANParks officials, have dismissed the Makuleke model on the grounds that it benefits few individuals and that the income generated from the operations in the park is often misused by the local elite. These weaknesses, in our view, are not strong enough to explain why the Makuleke model is not being replicated in other land claims in the KNP and elsewhere in the country's national parks such as Mapungubwe that had not been resolved. In a surprising but nevertheless bold move, the then two ministries of Agriculture and Land Affairs and Environmental Affairs and Tourism developed a common approach towards the resolution of land claims in protected areas in 2005. The approach was indeed finalised in the form of an inter-ministerial Agreement that was signed by the two ministries in 2007.

The key components of the Agreement are that the two ministries will adopt a unified and common approach to land claims in protected areas; no claimants, if successful, will be given settlement rights; the ecological integrity of protected areas will be secured; successful claimants will form an association that will represent them in co-management structures; the land restored can only be alienated to an organ of state and Broad-Based Black Economic Empowerment will be the channel for beneficiation (Memorandum of Agreement 2007). The Agreement makes it clear that any land lease or concession should be in line with the Protected Areas Act that upholds the integrity of all protected areas in the country. This way, the Agreement placed environmental protection firmly at the centre of land claims while leaving land claimants with some alternative options outside protected areas. When the Cabinet of the South African government approved this agreement in 2008, the option for resolving land claims the 'Makuleke way' was officially closed.

The Agreement that was approved by Cabinet has its supporters and detractors, with some groups torn between the two camps. Led by SANParks, supporters of this Agreement argue that national parks should not be restituted to any community in South Africa. To them, applying the Makuleke model to outstanding claims in the KNP will create a conglomeration of community-owned parts of the KNP; balkanize the park and complicate its management; render ecotourism economically unviable and compromise KNP's commercial viability to subsidise other national parks in the country (Senior SANParks Official, interview, 18 October 2011). In other words, communities as land owners in the KNP will precipitate a crisis. Supported by WESSA, SANParks suggested that outstanding land claims in the KNP should be settled by means of financial compensation. Critics of the Agreement see it as an environmental injustice; a way of sacrificing the victims of apartheid and unwilling-ness to give land claimants direct access to the market share of ecotourism (Mashile Mokono, interview, 12 February 2012). Conservationists who favour the Makuleke model are silent about why this model is abandoned, because any side they take would work against them. For example, any public support for the abandonment of the Makuleke model would nullify their earlier celebrations of the model. Supporting the use of the same model in other land claims in the KNP would be in conflict with their ideological position and vision for national parks as pristine wilderness. As we argued above, conservationists supported the Makuleke model, because the settlement of the land claim was in their favour at the time. Outstanding land claims have very little tangible benefits towards realising conservationists' goals in the country's protected areas. Instead, they appear as simply another nuisance!

Conclusion

Responses to outstanding land claims in the KNP reveal that conservationists who once celebrated the settlement of the Makuleke as a win-win solution are fiercely opposed to the model they help build. In fact, the attempt by the Makuleke to develop a third concession is seen by SANParks as a threat to the pristine wilderness and the ecological integrity of the Pafuri area that would turn the KNP into a tourism development zone (Senior SANParks Official, 18 October 2011). In our view, responses such as these reflect the commercial interests of SANParks and the concessionaires that are seen to be threatened by land claims. Clearly, the ecological argument is valid, but the timing of the shift away from the Makuleke model opens up other possible explanations, one of which is the realignment of conservation goals with business interests. Conservation is big business! As literature on the neoliberalisation of nature has shown, the alliance between capitalism and conservation takes different forms, and is contingent on the framing of environ-mental problems on which proposals for market-based solutions are anchored (Castree 2008; Igoe, Neves, and Brockington 2010; Büscher et al. 2012). That is to say the alliance between capitalism and conservation frequently, but not always, follows the problem-solution formulation that closes up spaces for alternative ways of, in the case of conservation, protecting nature. Whereas the initial solution to the problem of conservation in the KNP came in the form of ecotourism enterprises in

Makuleke, more land claims in the same vein challenge the monopoly of these enterprises.

The promotion of ecotourism as a solution to land claims is increasingly been questioned by both protagonists and critics of the Makuleke land claim settlement for different reasons. Protagonists question whether the Makuleke model is economically viable, and fear that the replication of such a model would fragment the KNP. For their part, critics of the model argue that it has placed severe limitations on the community and also over-estimate the benefits from ecotourism that are dependent on economic conditions. As ecotourism seems to be falling below the initial expectations, a new line of defence for the model is beginning to emerge. This line of thinking suggests that the Makuleke model should not be measured by the ecotourism enterprise associated with the land claim only. Instead, attention should also be given to other development projects in Makuleke village (Robins 2013). Read from this perspective, the land claim matters not so much because of the land rights it embodies, but its high value is to be found in its use as a marketing tool and a place-making process that presents Makuleke as a unique village. The uniqueness of the village acts as a magnet for forces of social change.

In summary, the experience with land claims in the KNP demonstrates that property rights can be supported or withdrawn momentarily in pursuit of particular business interests in nature conservation. We conclude that land claims aid our understanding of the symbiotic relationship between capitalism and conservation. While ecology stands out as the main reason for resolving land claims in the KNP, commercial interests played an equally important role in shaping the resolution of land claims. In the process, these commercial interests also influence land and agrarian relations in the sense that one group of people is allowed to reconstitute itself in relation to its ancestral land while others in the same situation are prevented from doing so.

Acknowledgements

The authors would like to thank the National Research Foundation for funding the research from which this paper draws material. Medupi Shabangu is thankful for a grant from the Charlotte Conservation Fellowship of the African Wildlife Foundation that supported the fieldwork.

Notes

1. The signatory to the agreement were: Makuleke community, SANParks, Minister of Environmental Affairs and Tourism, Minister of Public Works, Minister of Land Affairs, Minister of Minerals and Energy, Minister of Agriculture, Minister of Defence, and Member of the Executive Council for Agriculture, Land and Environment, Northern Province.
2. These are defined in the agreement as all activities, which are capable of being conducted within the Makuleke Region and are of an income producing or commercial nature; and which shall include but not necessarily be limited to ecotourism (Main Agreement Relating to the Makuleke Land Claim 1998).
3. We take this concentration of concessions in the KNP as an indication that the KNP was the main target of the commercialisation strategy.
4. This comment was made during questions and answers in a session on 'Land claims and nature conservation in South Africa' at the conference on Land Divided: Land and South African Society in 2013, March 24–27, Cape Town.

Notes on contributors

Maano Ramutsindela obtained his PhD in Geography at Royal Holloway, University of London, where he studied as a Canon Collins Scholar. He is currently Associate Professor in the Department of Environmental and Geographical Science at the University of Cape Town. He has researched and published widely on land reform in protected areas, borders, regions and transfrontier conservation. He is the author of *Parks and People in Postcolonial Societies* (Kluwer/Springer 2004, 2010); and *Transfrontier Conservation in Africa: At the Confluence of Capital, Politics and Nature* (CABI 2007). His latest book (with Marja Spierenburg and Harry Wels) is *Sponsoring Nature: Environmental Philanthropy for Conservation* (Earthscan/ Routledge, 2011). He can be contacted at: Maano.ramutsindela@uct.ac.za.

Medupi Shabangu is a Masters student in the Department of Environmental and Geographical Science at the University of Cape Town. His current research project is on contested resolutions of land claims in the Kruger National Park. He is conducting this research while at the employment of AgriSeta, where he is a project coordinator for rural development in Limpopo. His previous employment includes Community Development Officer (African Wildlife Foundation), and Chief Planner and Deputy Manager (Regional Land Claims Commission, Limpopo). He can be contacted at: shabangumedupi@gmail.com.

References

Arenstein, J. 1997. "Multi-Billion Dolphin Deal 'Top Secret'." *Mail and Guardian*, January 16. http://mg.co.za/article/1997-01-10-multi-billion-dolphin-deal-top-secret.

Bakker, K. 2005. "Neoliberalizing Nature? Market Environmentalism in Water Supply in England and Wales." *Annals of the Association of American Geographers* 95 (3): 542–565. doi:10.1111/j.1467-8306.2005.00474.x.

Bond, P. 2000. *Cities of Gold and Townships of Coal: Essays on South Africa's New Urban Crisis.* Trenton, NJ: Africa World Press.

Brockington, D., and R. Duffy. 2010. "Capitalism and Conservation: The Production and Reproduction of Biodiversity Conservation." *Antipode* 42 (3): 469–484. doi:10.1111/j.1467-8330.2010.00760.x.

Brockington, D., and J. Igoe. 2006. "Eviction for Conservation: A Global Overview." *Conservation and Society* 4 (3): 424–470. http://www.conservationandsociety.org/text.asp? 2006/4/3/424/49276.

Brockington, D., and K. Scholfield. 2010. "The Conservationist Mode of Production and Conservation NGOs in Sub-Saharan Africa." *Antipode* 42 (3): 551–575. doi:10.1111/j.1467-8330.2010.00763.x.

Bryant, R. L. 2009. "Born to be Wild? Non-Governmental Organisations, Politics and the Environment." *Geography Compass* 3 (4): 1540–1558. doi:10.1111/j.1749-8198.2009.00248.x.

Büscher, B., S. Sullivan, K. Neves, J. Igoe, and D. Brockington. 2012. "Towards a Synthesized Critique of Neoliberal Biodiversity Conservation." *Capitalism Nature Socialism* 23 (2): 4– 30. doi:10.1080/10455752.2012.674149.

Carrier, J. G. 2010. "Protecting the Environment the Natural Way: Ethical Consumption and Commodity Fetishism." *Antipode* 42 (3): 672–689. doi:10.1111/j.1467-8330.2010.00768.x.

Carruthers, J. 1995. *The Kruger National Park: A Social and Political History.* Pietermaritz-burg: University of KwaZulu-Natal Press.

Castree, N. 2008. "Neoliberalising Nature: The Logics of Deregulation and Reregulation." *Environment and Planning A* 40 (1): 131–152. doi:10.1068/a3999.

Consultancy Africa Intelligence. 2011. "Land Restitution, Community Rights and Conservation: A Success Story (Discussion Paper/Optimistic Africa)." Accessed December 15, 2012. http://www.consultancyafrica.com/index.php?option=com_content&view=article&id=841: land-restitution-community-rights-and-conservation-a-success-story&catid=90:optimistic-africa&Itemid=295

Ellis, S. 1994. "Of Elephants and Men: Politics and Nature Conservation in South Africa." *Journal of Southern African Studies* 20 (1): 53–69. doi:10.1080/03057079408708386.

Ferrar, T., A. Wynne, and M. Johnstone. 1997. *Community Based Tourism in the Northern Province*. Johannesburg: Land and Agriculture Policy Centre.

Gray, A. 1998. *The Dolphin Deal: The Real Issues Behind the 'Controversy'!* Unpublished document, 17 March.

Harthon, M. 1998. *Minutes of the Meeting on Mineral Rights and Mining Issues*. Pretoria.

Harvey, D. 2005. *A Short History of Neoliberalism*. Oxford: Oxford University Press.

Heynen, N., J. McCarthy, S. Prudham, and P. Robbins, eds. 2007. *Neoliberal Environments: False Promises and Unnatural Consequences*. London: Routledge.

Igoe, J., K. Neves, and D. Brockington. 2010. "A Spectacular Eco-tour Around the Historic Bloc: Theorising the Convergence of Biodiversity Conservation and Capitalist Expansion." *Antipode* 42 (3): 486–512. doi:10.1111/j.1467-8330.2010.00761.x.

Kepe, T., R. Wynberg, and W. Ellis. 2005. "Land Reform and Biodiversity Conservation in South Africa: Complementary or in Conflict?" *International Journal of Biodiversity Science and Management* 1 (1): 3–16. doi:10.1080/17451590509618075.

King, B. 2009. "Commercializing Conservation in South Africa." *Environment and Planning A* 41 (2): 407–424. doi:10.1068/a4016.

Koch, E. 2003. "Summary of Anticipated Makuleke Payments." Paper presented at the World Parks Congress, Durban, September 8–17.

Magome, H., and J. Murombedzi. 2003. "Sharing South Africa's National Parks: Community Land and Conservation in a Democratic South Africa." In *Decolonizing Nature: Strategies for Conservation in a Post-colonial Era*, edited by W. M. Adams and M. Mulligan, 108–134. London: Earthscan.

Main Agreement Relating to the Makuleke Land Claim. 1998. *Agreement between the Makuleke Community, Ministers and the South African National Parks*. Unpublished document. Pretoria.

MacDonald, K. I. 2010. "The Devil is in the (Bio)Diversity: Private Sector 'Engagement' and the Restructuring of Biodiversity Conservation." *Antipode* 42 (3): 513–550. doi:10.1111/j.1467-8330.2010.00762.x.

Mavhunga, C., and M. Spierenburg. 2009. "Transfrontier Talk, Cordon Politics: The Early History of the Great Limpopo Transfrontier Park in Southern Africa, 1925–1940." *Journal of Southern African Studies* 35 (3): 715–735. doi:10.1080/03057070903101920.

McAFee, K. 1999. "Selling Nature to Save it? Biodiversity and Green Developmentalism." *Environment and Planning D: Society and Space* 17 (2): 133–154. doi:10.1068/d170133.

Memorandum of Agreement. 2007. Memorandum of Agreement between Ms Lulama Xingwana (Minister of Agriculture and Land Affairs) and Mr Marthinus van Schalkwyk (Minister of Environmental Affairs and Tourism). Pretoria.

Mpumalanga Parks Board. 1998. "Community Leaders Support Commercialization of Loskop Dam Nature Reserve." *Press Release*, June 29. http://www.mtpa.co.za/index.php?parks+1812.

Nash, R. F. 2001 *Wilderness and the American Mind*. 4th ed. New Haven, CT: Yale University Press.

Ramutsindela, M. 2002a. "The Perfect Way to Ending a Painful Past? Makuleke Land Deal in South Africa." *Geoforum* 33 (1): 15–24. doi:10.1016/S0016-7185(01)00008-2.

Ramutsindela, M. 2002b. "The Globalisation of South Africa's Natural Capital." In *South Africa since 1994: Lessons and Prospects*, edited by S. Buthelezi and E. Le Roux, 73–91. Pretoria: Africa Institute.

Ramutsindela, M. 2007. *Transfrontier Conservation in Africa: At the Confluence of Capital, Politics and Nature*. New Haven, CT: Cabi. doi:10.1079/9781845932213.0000.

Ramutsindela, M., M. Spierenburg, and H. Wels. 2011. *Sponsoring Nature: Environmental Philanthropy for Conservation*. London: Earthscan.

Robins, S. 2013 "The Elephant in the (Lowveld) Room? Eco-tourism-led Development and the Commodification of Nature in Makuleke Village, Limpopo Province." Paper presented at the conference on Land Divided: Land and South African Society in 2013, Cape Town, March 24–27.

South African Government. 1996. *Government Notice 1064*. Pretoria: Government Printer.

South African Government. 1997. *White Paper on Conservation and Sustainable Use of South Africa's Biological Diversity*. Pretoria: Government Printer.

South African National Parks. 2001. *Concession Contract*. Pretoria: South African National Parks.

South African National Parks. 2007. *Annual Report, 2006/7*. Pretoria: South African National Parks.

Spierenburg, M., C. Steenkamp, and H. Wels. 2006. "Resistance of Local Communities against Marginalization in the Great Limpopo Transfrontier Conservation Area." *Focaal* 2006 (47): 18–31. doi:10.3167/092012906780646479.

Treaty of the Great Limpopo Transfrontier Park. 2002. *Treaty between the Government of Mozambique, the Government of the Republic of South Africa and the Government of Republic of Zimbabwe on the Establishment of the Great Limpopo Transfrontier Park*. Pretoria.

Varghese, G. 2008. "Public-private Partnerships in South African National Parks: The Rationale, Benefits and Lessons Learned." In *Responsible Tourism: Critical Issues for Conservation and Development*, edited by A. Spenceley, 69–83. London: Earthscan.

Markets of exceptionalism: peace parks in Southern Africa

George Barrett

Department of Political and International Studies, Rhodes University, Grahamstown, South Africa

The vision of Southern African peace parks – transfrontier conservation areas – is one of 'boundless' natural landscapes transcendent of the brutality of the cartographic legacy of sovereign-statism. The parks are presented as vast biodiversity rich wildernesses inhabited by rare and precious fauna and flora and scattered communities of 'traditional' African peoples. As such, the region's frontiers symbolise exceptional spaces for ecosystem scale conservation, and the fostering of peace, community, and economic prosperity in a historically troubled region. However, the peace parks vision is emblematic of the commodification of life that pervades strategies for environmental governance and conservation in the current neoliberal era. Re-reading the vision through discourses of 'the exception' and securitisation demonstrates how its partiality and performativity act to discipline the region's borderzones in line with market priorities. In doing so, the vision (re)creates historically rooted patterns of inclusion and exclusion, security and insecurity in the life of the parks.

The borderzones of Southern Africa have, and continue to be, spaces of exception where the adoption of extraordinary measures have been legitimated in response to perceived threats to state security.[1] The orthodox readings of International Relations (IR) place significance on the function of borderzones as the demarcation of sovereign territorial power. Despite the Organisation of African Unity's acceptance of the colonial carving up of the continent (OAU Charter, OAU 1981), the arbitrary nature of the borders means they are sites of continual contestation and the assertion of state authority in the defence of the 'nation'[2] and national security. However, during colonialism and apartheid,[3] the nation was invariably defined by those with the power and authority to do so and the borderzones became spaces of both physical and symbolical exclusion. They were thus zones of exception where the state, through a combination of bureaucratic checks and armed patrols, was able to exercise its authority. Although not universally successful, processes used to discipline these spaces were legitimated by the primacy given to national security and the silencing of related insecurities, dislocation and alienation experienced by the resident and transient communities of the region's borderzones.

However, in the contemporary era, the state is only one of a plethora of actors able to declare and frame the 'exception' in response to a variety of threats to a range of referents. Consequently, a 'field of security relations, between security professionals, governmental and non-governmental institutions, the police, military, and private enterprise, across an increasingly *globalised* terrain' has emerged (Peoples and Vaughan-Williams 2010, 69, emphasis in the original). Attention must therefore be given to the host of 'transnational networks of bureaucracies and private agents who "manage" the (in) security underpinning the exception' (Bigo 2008, 116) in relation to a specific referent and context. Given the neoliberal capitalist forces underpinning contemporary modernity, attention must also focus on the structures and processes of consumption, the routines of which frame 'the condition of possibilities of these claims and their acceptance' (128).

In post-apartheid Southern Africa, advocates of peace parks – transfrontier conservation areas – have constructed these borderzones as exceptional spaces of an alternative kind. A consortium of public, private, and state sponsors argue that the development of peace parks transcends the brutality of this cartographic legacy to become spaces of opportunity for the realisation of a host of national, regional and global environmental, socioeconomic, and political goals. These actors view the 'progress promised by the state as the best form of protection' and use this promise to 'inculcate the region's life with [particular] values of modernity' (Vale 2003, 165). In the current era of neoliberalism, this entails 'the transformation of previously untradeable things and ideas into commodities that are visible and tradable in the world capitalist economy' (McAfee 1999; Castree 2007 cited in Brockington, Duffy, and Igoe 2008, 191). Thus, although elements of neoliberalism may appear to undermine state authority, the state continues to be a major vehicle through which other actors can pursue their own objectives and in so doing, bolster the state in the process. Consequently, transnational networks compete to frame and prioritise the perceived threat agenda, the course of action required to address it, the mechanisms necessary to maintain the security of the referent, and to defeat or contain the elements causing the threat (Bigo 2008, 128). However, they may do so independently of, or in collaboration with, the state.

This paper argues that the vision of peace parks and the processes required to secure them create patterns of insecurity through the routinisation of illiberal practices of inclusion and exclusion that result from the commodification of wildlife, landscape and people in the borderzones of Southern Africa. Furthermore, it argues that these practices of inclusion and exclusion are often obscured or ignored by the focus on the 'spectacle'[4] embodied in the vision of peace parks and what it claims to offer in terms of addressing these global (environmental) priorities, including the conservation of the world's ecological integrity, the sustainability of the neoliberal economic order itself, and the pursuit of a more secure and prosperous future for Southern African states and their people (Barrett 2011).[5]

The promise of peace parks in an era of neoliberal conservation

The rich and varied biodiversity of Southern Africa has made peace parks an appealing option through which to contribute towards multiscalar environmental rescue through the conservation and protection of charismatic and rare flora and fauna including: elephants, rhinoceroses, large cat species, and the endangered *ansellia Africana* (leopard orchid). In addition, they contribute to ecosystem scale conservation and, more recently, carbon sequestering projects (PPF 2006, 23).[6] The high profile of environmental challenges in the contemporary era gives both urgency and legitimacy to the parks creation as one in a plethora of possible 'solutions'. Southern African peace parks are also perceived by their advocates as the vehicle for the delivery of a host of additional benefits. At the local and regional level the parks have the potential to counter, in part at least, the legacy of maldevelopment[7] and structural poverty created through colonial and apartheid rule in the borderzones and at the national level. They can do this through the growth of public, private and community eco and cultural tourism ventures, increased visitors to respective national parks and the expansion of ancillary business sectors. As such, the parks can potentially contribute to the reconstruction of the region's image, and help to overcome the historically rooted regressive constructions of Southern African states, particularly South Africa, as envisaged in Thabo Mbeki's African Renaissance. In turn this may help spur further economic growth and investment.[8] As well as contributing to a host of economic and environmental 'goods', the idea of peace parks is also to foster interstate peace and cooperation in a historically troubled and divided region (Sandwith et al. 2001).[9] Given the arbitrary nature of the borders and their historically contested nature, the borderzones designated as peace parks can be understood as an exceptional response to a complex interplay of national, regional and global environmental, economic, and political concerns.

The growing trend towards so-called eco-friendly or green consumerism, eco-corporate social responsibility initiatives, and 'natural' capital investment opportunities are all manifestations of prevailing global environmental governance that have succumbed to the logic of neoliberalism (Sandilands 1993; Igoe and Brockington 2007; Büscher and Dressler 2007; Brockington, Duffy, and Igoe 2008; Carrier 2010; Büscher 2010b). Market-based solutions to environmental degradation and destruction have therefore, become uncritically accepted and normalised within mainstream conservation practices (MacDonald 2010), particularly those 'encouraged' in the developing world in relation to protected areas (Igoe 2010; Igoe, Neves, and Brockington 2010). Put differently, the need to 'secure' our ecological and economic future has resulted in the global shift towards the commodification of the environment and elements thereof (Igoe and Brockington 2007; Brockington, Duffy, and Igoe 2008). Tied to the worldwide expansion of tourism, and more specifically ecotourism, the commodification of the environment carries the promise of economic growth and employment opportunities through both state and private ventures. This is particularly appealing in Southern Africa, which experiences high levels of unemployment and stunted economic growth. Southern (and Eastern) African states thus seek to recreate the colonially rooted 'spectacle' of exclusive 'African wilderness' to attract high-paying foreign visitors. This is enhanced by the idea of a boundless African landscape encapsulated in the notion of peace parks (Dressler and Büscher 2008). Consequently, there has been a degree of synergy among the region's states, the Southern African Development Community (SADC), and key conservation actors, including the Peace Parks Foundation (PPF), World Wildlife Fund (WWF) (South Africa), and the International Union for the Conservation of Nature (IUCN), in the construction of a marketable vision of the parks that plays significantly on their apparent exceptionalism.

Peace parks, the PPF and Boundless Southern Africa (BSA)

The development of peace parks, Southern African peace parks in particular, is well documented (Ali 2007; Hanks 2003; PPF *Origins* n.d.; Thorsell 1991; Westling 1993; Wolmer 2003). In brief, peace parks are not unique to the region, and when first tabled by the South African Afrikaner business tycoon and conservationist, Anton Rupert, to the then President of Mozambique – Joaquim Chissano – in May 1990, they were already established as an appropriate biodiversity conservation option in international environmental governance circles (PPF Origins n.d.). Rupert, then President of the Southern African Nature Foundation (now WWF South Africa), sought to 'recreate' the historic migration routes for the region's wildlife disrupted by political state boundaries by linking 'islands' of biodiversity through a series of vast transfrontier parks and corridors across Southern Africa. Amended in the early 1990s by the World Bank's Global Environmental Fund (GEF) to include socioeconomic development objectives, Rupert's dream was first realised with the official opening of the Kgalagadi Transfrontier Park (TFP) between South Africa and Botswana in 2000 (SADC n.d.). The PPF website lists 10 Southern African peace parks, with 22 potential sites identified across the region (Myburgh n.d.). The largest, the Greater Limpopo Transfrontier Park, covers 35,000 km^2, and is part of the much larger Greater Limpopo Transfrontier Conservation Area covering a staggering 100,000 km^2. However, this is set to be succeeded by the Kavango-Zambezi (Kaza) Transfrontier Park following the Memorandum of Understanding (MoU) signed in 2006 between Angola, Botswana, Namibia, Zambia and Zimbabwe, which will potentially cover an estimated 287,132 km^2 (Fox 2009). Each park is recognised for its particular biodiversity and landscape, as well as the unique ecotourism experience it can offer including a range of adventure and 'cultural' experiences.

Founded by Anton Rupert, the PPF has since its inception positioned itself as *the* coordinating body in the establishment and construction of Southern African peace parks. Benefitting from the profile of its visionary – Rupert – the PPF has proven adept at drawing in huge amounts of donor funding and the support of high profile political figures, including Nelson Mandela and Prince Bernhard of the Netherlands (PPF *Origins* n.d.). It has also been pivotal in securing the continued support of Southern African states – although sometimes more in rhetoric than practice – and international 'approval' for the development of the parks. Working with regional states, the PPF appear to have the 'social and symbolic capital' required to convince its external and state level audiences of the viability and necessity of the parks (Spierenburg and Wels 2010; PPF *Donors* n.d.).[10] This is evidenced not only by the number of parks in various stages of development, but also by the PPF's authority as part of a wider network of conservation actors in influencing government policy on key areas of concern.[11] In contrast, BSA is an inter-state collaborative marketing strategy aimed at promoting the peace parks vision to potential tourists and investors. In particular, the *BSA* website showcases a 2009 Expedition that celebrates the peace parks vision and the unique cultural and wilderness experience it offers.

The PPF and BSA vision

The literature relating to the parks and particularly that of the PPF and *BSA* emphasises the parks' exceptionalism for ecosystem scale biodiversity conservation (see, for example Brosius and Russell 2003; Büscher 2010a; Schwartz 1999). This is a practice currently privileged in international environmental governance strategies promoted by leading conservation organizations, as well as global financial institutions, such as the World Bank and the International Monetary Fund (IMF) through its Global Environmental Facility (GEF) (Duffy 2005; SANParks n.d.; Spenceley 2006). These actors play a central role in framing global environmental governance and have been particularly vigorous in their funding of climate change and biodiversity conservation projects in the 'majority world'.[12] However, their strategies are underpinned by a pervasive, but selective understanding of conservation biology and environmentalism that has emerged in an era increasingly defined by the global advance of neoliberalism (Duffy 2007, 60; Igoe and Brockington 2007; Büscher and Dressler 2007; Büscher 2010b).

As a result, contemporary environmental strategies have come to embrace the commodification of nature, 'wilderness', and natural landscapes as the means to their salvation, not only for the continued benefit of humanity, but also for the long-term stability and growth of the neoliberal order (See, for example, King and Stewart 1996; West, Igoe, and Brockington 2006; Büscher and Dressler 2012). It is, after all, in wilderness that the 'preservation of the world' lies (David Thoreau cited in Cronon 1996). In particular, they embrace the politics of the spectacle (Debord [1967] 1995), because it is this, which makes cultural and natural entities commercially desirable. Thus, the parks are understood to be exceptional; in scale and grandeur, in the political and economic collaboration they foster, in the rarity of the flora and fauna and the landscapes they encapsulate, and in the nature of the transfrontier adventures that one can experience in these few remaining and seemingly unspoiled 'wildernesses'. However, as Cronon (1996, 7) argues, wilderness is 'quite profoundly a human creation'.

The spectacle has two key elements. First is the showcasing of the aesthetic beauty of the wildlife and landscape, and the wealth and variety of biodiversity interwoven with a sense of mythical and historic splendour, and the 'traditional' culture(s) of indigenous communities. Second is the celebration of projects undertaken by the PPF and BSA that give life to the vision on the ground. These include, for example, the 2009 BSA Expedition, the!Ae!Hai Kalahari Heritage Park, community development projects in and around the Maputo Special Reserve in the Mozambican side of the Lumbobo TFCA, and in the training of wildlife 'managers', field guides and hospitality personnel (PPF *Training Colleges* n.d.). They represent progress in meeting the dual conservation and socioeconomic development needs, but are only possible if the spectacle is successfully secured so that it can continue to attract consumers and investors. The excerpts below are illustrative of how the spectacle of an exceptional boundless wilderness and 'traditional' cultures are created in the PPF and BSA literature as part of the overarching peace parks vision. It is this vision, rather than the actual nature, that is, Büscher (2010a, 261) argues, more important for stimulating investment but also for specific constructions of how people and nature 'ought to be'.

Through their website, campaign videos and annual reports, the PPF emphasises the aesthetic beauty and richness of the 'timeless' natural scenery, wildlife and cultural diversity of the Southern African landscape. It is not only primarily in the textual and photographic representations of the parks, but also in the reportage of the PPFs related activities that the spectacle comes alive. The potential jewel of the peace parks, the Kavango-Zambezi, embraces the 'largest inland delta in the world, the biggest transfrontier conservation area in the world, breathtaking falls, and the largest contiguous population of African elephant and the highest concentrations of wildlife on the African continent' (BSA *Kavango-Zambezi* n.d.). The Maloti-Drakensburg TFCA hosts 'the highest falls in Southern Africa, home of the *critically endangered* bearded vulture' (BSA *Kavango-Zambezi* n.d., Emphasis added). The Greater Limpopo TFCA is described as the 'world's greatest animal kingdom'; a phrase subsequently repeated in the *BSA* campaign and in related media coverage (PPF *Great Limpopo* n.d.; MoEAT 2001; AWF 2002). In addition, the Lumbobo TFCA and Resource Area between Mozambique, South Africa and Swaziland, has 'one of the most striking areas of biodiversity' (PPF *Lumombo* n.d.). However, it is the Kgalagadi and the Ai/Ais Richtersveld which speak to the idea of a 'boundless wilderness' most directly. The Ai/Ais Richtersveld spans 'some of the most spectacular arid and desert mountain scenery' in Southern Africa, featuring the world's second largest Canyon, that of the Fish River (PPF *Ai/Ais/*n.d.). The Kgalagadi, a landscape relatively free from 'human interference',[13] is the 'the largest expanse of continuous sand mass in the world .... A remarkable place of shimmering heat and sand that seemingly stretches beyond the horizon' (BSA *Kgalagadi* n.d.). More explicitly a marketing campaign, BSA invites its audience to be 'mesmerised' by the Kgalagadi's scenery and 'stare down at ancient history weathered by an eternity of sun, wind and strong river currents' (BSA *Kgalagadi* n.d.). The exceptionalism of the wildlife is reinforced through reference to the 'world's fastest animal on land', 'the world's heaviest flying bird', and the 'unique Kalahari Lion' (BSA *Kgalagadi* n.d.).

This exceptionalism is matched by the representation of 'traditional' communities and their ancestors within the spectacle of the parks. The !Ae!Hai Kalahari Heritage Park within the Kgalagadi TFCA, for example, was established by South Africa's National Parks (SANParks) from community leased land to 'preserve the cultural and traditional knowledge of the historically marginalised ‡Khomani San and Mier communities, whilst improving their livelihood opportunities' (PPF *!Ae!Hai Heritage Park* n.d.). 'Intrepid adventurers' are invited to climb down the '350 million-year-old and erosion-rich Orange River gorge' abundant with 'history, folklore and grandeur' to 'touch the passage of time' (BSA *Ai/Ais/*n.d.). Tourists are also encouraged to visit the Imbewu Camp, where the traditional veld school for teaching the Bushmen children the way of their ancestors is located (PPF *!Ae!Hai Heritage Park* n.d.). In particular, it is the traditional lifestyles of the Nama in the Ai/Ais Richtersveld and the San of the Kgalagadi that are being '*preserved*' (PPF *Ai/Ais/*n.d. Emphasis added) as part of the spectacle.

Reference to pre-colonial times, to ancient African civilisations in the spirit of Thabo Mbeki's (2004) African Renaissance[14] is central to the spectacle being created. A visit to the 'cultural' TFCA, that of Greater Mapungubwe, will enable those seeking to 'experience a kinship with past generations' to explore Iron Age sites on all sides of the Botswana-South Africa-Zimbabwe borders. These sites, it is claimed, represent 'a highly sophisticated civilisation, which traded with Arabia, Egypt, India, and China' (PPF *Greater Mapungubwe* n.d.). The PPF therefore considers the Mapungubwe Heritage site to be a 'milestone' in the realisation of the African Renaissance and the rewriting of African history (PPF, *News: World Heritage Site*, July 9, 2003). The Maloti-Drakensberg TFCA is also highlighted for its cultural significance as the 'world's greatest outdoor gallery' and 'largest and most concentrated group of [San] rock paintings' south of the Sahara (PPF *Maloti-Drakensberg* n.d.). Implicitly, the bonds between these historic communities are invoked as transcendent of the modern international borders, which have hindered the unity of their decedents in recent centuries. Yet, through the spectacle this history is given life and continuity in the ‡Khomani San, Mier and Nama communities of today. An entry from the BSA Expedition diary is illustrative of the point:

> At sunset we walk across the pan to meet these delightful San people, like early hunter gatherers we squat around the coals of a small fire on the Kalahari sand where amongst rudimentary grass and stick shelters, we watch the light golden brown complexioned women with smiling wrinkles and twinkling laughing eyes, using their tiny hands to fashion ostrich shell necklaces and bangles. (Gertruida 2009)

Such descriptions speak to BSA's slogan of 'Nature, Community and Culture' as part of the spectacle and thus, consumer experience. In a promotional video, images of the landscape and wildlife are set to a soundtrack of spirited music and interspersed with title pages, one of which reads, 'where communities and cultures are supported and celebrated' (BSA 2008). The punctuation of the narrative with high quality photography profiling the stunning cultural and natural beauty of the parks and their inhabitants gives further visual expression to the spectacle being constructed.

This is evidenced further in the video recording of the performance of the 'Boundless Song' by singers in traditional cultural dress inviting tourists to visit 'boundless' Southern Africa and experience first-hand the region's wonders; to celebrate the freedom with which animals roam and people collaborate (BSA *Boundless Song* n.d.).[15] The PPF argues that the 'national boundaries proclaimed at the Treaty of Berlin in 1884 . . . cut across tribal and clan groupings, as well as wildlife migration routes, fragmenting ecosystems and *threatening* biodiversity, while the establishment of peace parks strives to correct these past injustices' (PPF *Why Peace Parks?* n.d. Emphasis added). The 'opening' of migratory routes to enable the region's wildlife to roam freely, as they have 'since the dawn of time' (PPF *Lower Zambezi- Mana Pools* n.d.) is a symbolic transcendence of this violent history, and one which is again commonly referenced in the media progress reports of the parks' developments.

The sensory construction of the peace parks vision is not, however, necessarily sufficient either to sustain consumer and investor attention nor the parks' legitimacy as an extraordinary but crucial response to the conservation and development challenges for which they were designed. Thus, the vision requires constant articulation and performativity[16] for it to become a marketable and tangible 'reality'. Performativity is understood in the Weberian sense, as 'the ongoing citational processes whereby "regular subjects" and "standards of normality" are discursively constituted to give the effect that both are natural rather than cultural constructs' (Weber cited in Dunn 2009, 431). Thus, the vision of the parks is 'performed' in order to make it appear natural. This performativity is achieved through a series of discursive practices at a range of levels – from the state down to the daily functioning of the park – which are then promoted as 'milestones', 'progress' or 'success' stories. This is explicitly evident in the PPF's reporting on its 'investment' in community and sustainable livelihoods initiatives, through community collaboration in the Joint Management Boards (JMBs) with respective national park representatives, and through wildlife management and leisure industry training. All of which speaks to the apparent enthusiasm of community members to 'join the wildlife business' (PPF *Lumombo* n.d.).[17] This is based on the claim that 'all affected communities as stakeholders identified their needs and priorities for development' as part of 'integrated development plans (IDPs)' managed by the PPF (PPF *Kavango-Zambizi* n.d.).[18] The PPF also speaks of efforts to 'manage' human–nature conflict by, for example, providing communities with electric fences to protect their crops from marauding elephants (PPF *Greater Mapungubwe* n.d.).

At the inter-state level the publicised celebrations surrounding the signing of Memorandums of Understanding (MoU) or treaties between collaborating states gives tangible expression to the parks. They help to validate the PPF claims that the parks are 'an African success story', representing an 'exemplary process of partnership between governments and the private sector' and where 'peace reigns between the relevant countries' as evidenced by the 'free movement of tourists and wildlife across international borders' (PPF *Southern African Peace Parks?* n.d.). Thus, the creation of the parks, even just on paper, helps to underpin the vision and itself represents a remarkable achievement for a region with such a turbulent inter-state history.

Once created, tourists can experience the 'wildernesses' and rich biodiversity of the region first hand. To entice them, a range of activities are promoted, including the possibility of 4×4 trails to experience, for example, the 'Kalahari's tranquillity' (PPF *Kgalagadi* n.d.). When this becomes a genuine possibility, a lived experience for those visiting the parks, it helps to solidify the vision by making it 'real' and this is where the reportage of the 2009 BSA Expedition comes into its own. The Expedition, led by Kingsley Holgate, one of Africa's 'most colourful modern day explorers' (Holgate 2011) traversed the continent from the Indian Ocean to the Atlantic. It linked '9 Southern African Countries, 7 Transfrontier Conservation Areas, more than 30 Game and Nature Reserves and the communities living in and adjacent to these areas' as documented in detail on the BSA website (BSA 2009). The aim was to 'boost Southern Africa's magnificent parks...removing the barriers that have not only prevented communities from benefiting from resources in the area, but also blocked the natural migration routes of wildlife' (MCSA 2009). 'Mad Mike', one of the Expedition team-members, said that it had 'helped keep the Peace Parks Foundation's vision of Transfrontier Conservation alive, mapped, photographed, filmed, documented, raised awareness for and opened up what...has the potential to become accepted internationally as one of the greatest coast-to-coast 4×4 adventure routes in the world' (Rumble n.d.).

The tag line of 'open spaces, unlimited beauty, [and] infinite possibilities', is continually reinforced in the route maps and diary entries of the expedition team available on the BSA website. The diary entries give life to the sense of adventure, of the spectacle and of the unity of 'nature, community, and culture' as this extended extract illustrates:

> *We're still zigzagging across Mama Africa, and she's beautiful as ever, as now out of Caprivi and back into beautiful Botswana, wild dogs at Savuti, we head for the River Khwai. The road is flooded, difficult going, abnormally high water, deep river crossings – we camp on the riverbank.... Next day.... – another great community day to link nature, culture and community. Speeches in the 'Kgotla', the chief's meeting place, the gathering of Khwai River water to be added to the symbolic expedition calabash, this time by two lovely traditional Basarwa San ladies. There's singing, dancing, feasting...Onto Xakanaxa in*

Moremi Game Reserve – more flooded river crossings, elephants galore, a great leopard sighting – no. 6 on this journey – the area is always a paradise for wildlife. Ross' voice over the radio: "Another leopard across the mopani pole bridge under the sausage tree." That's no. 7 – let's face it, the freedom of a 4x4 wildlife journey across Botswana is a must for every adventurer – a hardwood fire at night, the unfenced sounds of the wild, the roar of a lion, red-billed francolins heralding the dawn, the cackle of a hyena, the cry of a jackal, and the true Transfrontier travellers, the ever present elephant who need no boundaries – some 250,000 of these silent giants that wander across the Kavango Zambezi Transfrontier Park, shared by Angola, Botswana, Namibia, Zambia and Zimbabwe. (Gertruida 2009)

The extract is one of many reinforcing the idea of a BSA being actually experienced. The Expedition thus adds a different dimension to the performativity of the peace parks vision. Supporting such ventures, better still, actively participating in them, celebrates the perceived role ecotourism has in uplifting marginalised, rural communities out of poverty; poverty which otherwise causes them to 'exhaust' the resources upon which 'their and *our* survival depends' (Rabson Dhlodhlo cited in Blandy 2006. Emphasis added). As the PPF's former Chief Executive, Prof. Willem van Riet stated, 'with the worldwide growth in the ecotourism industry, peace parks offer an opportunity to optimise the abundance of fauna and flora that Africa has to offer – to the benefit of local communities' (cited in Blandy 2006). BSA and the PPF therefore, offer consumers and investors an insight into what might be referred to as, to employ a contemporary cliché, the 'feel good factor' of their investment. In so doing, they reinforce the appeal of the commodification of nature in conservation (Büscher and Dressler 2012) as the source of its salvation and the alleviation of poverty and mal/underdevelopment. The success of the vision created and its performativity thus makes critique of the neoliberal order underpinning it – which is the cause of so much environmental and socioeconomic ills – difficult. Furthermore, it brings the borderzones of Southern Africa firmly under the direct influence of a powerful consortium of state and private elite actors who champion this very order.

However, constant articulation of the spectacle, and the positive performativity of it, is insufficient for securing the vision. As the very idea of a 'park' suggests, these are highly managed spaces, despite the impression created in the marketing material that suggests otherwise. Thus, the vision and its manifestation give rise to, and are dependent upon, the successful disciplining of the life of the parks in conformity with the vision. This disciplining is either completely absent, obscured, or euphemised in the marketing literature. Moreover, it is normalised in the consumer experience of the parks, because it visibly manifests in the ritual identity checks of international travel and the familiar presence of security personnel in daily life. So, for example, visitors rarely stop to question who, by virtue of possessing or not possessing the right documentation let alone the socioeconomic means, is permitted entry into these 'boundless' spaces. The routinisation of such checks are further legitimated by the urgency of the environmental and development challenges the parks seek to address and the promise they hold to do so; and reinforced by the authority of the actors constructing the vision. It is, however, primarily because of the power of the vision itself and the pervasive logic of the market which draws on the spectacle to grow and expand through the ever increasing commodification of life. To reiterate, it is this particular manifestation of neoliberalism that frames the 'the condition of possibilities' (Bigo 2008, 128) of the vision and its acceptance.

Marketing routines of exclusion and insecurity

It is in this sense that the vision can be understood to operate like the 'exception', giving rise to extraordinary measures of discipline and control in the context of the borderzones. These measures appear both logical and desirable in defence of the vision and its realisation. However, for those they render insecure, for those they alienate and marginalise and whose lives they control, and for those upon whom they re-inscribe regressive identity constructions, they appear illiberal. Moreover, they impact on people invariably already marginalised within society and located within spaces that have historically been at the periphery (of the state). However, the disciplining processes take multiple forms, as discussed in more detail below. Firstly, they are in the very construction of the spectacle as detailed above. Secondly, they are prevalent in the privileging of 'rational' scientific and technological processes of understanding that underpin global priorities for conservation and development strategies, which in turn inform the structure and designation of land-use within the parks and buffer zones. Thirdly, they are in the physical and administrative policing of the parks. Operating in unison, these processes render the lived life of the parks more predictable, efficient, and controllable. Yet, they also embody the 'politics of inclusion and exclusion' embedded in neoliberal structures of prevailing 'global governmentality' (Tosa 2009, 414).[19]

The PPF and BSA's construction of an African wilderness; one relatively untouched by people, is contradicted in several ways that express the exclusionary nature of the vision and the means of its security. For example, the expedition encourages adventurers to 'flock' to the Kalahari; the realisation of which would, of course, reduce the exceptionalism of the spectacle and its successful commodification. Neither the PPF nor BSA are targeting mass tourism, it is the relative exclusivity of the experience that makes it so commercially desirable. Thus, the experience on offer is explicitly targeted at elite travellers who have the luxury of time and money to embark on such an adventure. Not only is it therefore, exclusionary in socioeconomic terms, but as Hughes (2002, 2) argues, invariably also in racial terms – given the regional distribution of wealth, particularly in South Africa – from where the majority of regional tourists originate. It is also reinforced by the explicit declaration by the Botswanan government and others – supported by the IUCN and international donors – to target high-end tourists as part of their national tourism strategy (See Stevens and Jansen 2002). The vast majority of the region's people are thus automatically excluded from the enjoyment of the spectacle within their own countries.

There is also a dualism in the construction of the peace parks vision that is obscured in the promotional material. On the one hand, global environmental governance strategies for biodiversity conservation 'best practice' have been used to legitimate ecosystem scale conservation initiatives. This 'best practice' exemplifies a move away from community-based natural resource management (CBNRM) and a return to 'fortress conservation' reminiscent of colonial practices, albeit more recently with a pro-community veneer (Dressler and Büscher 2008). On the other hand, and in an era where human and minority rights have gained substantial international attention, states are arguably more sensitive to demonstrating recognition of and support for these communities. This is made more appealing when indigenous communities and their cultural 'traditions' can become revenue earners. To this end, BSA and the PPF celebrate, as part of the spectacle, the indigenous peoples that have struggled for the right to remain in these spaces, including those of the Nama, ‡Khomani San and Mier.

These communities are constructed as symbolic of the unity of 'nature, community and culture' explicitly invoked in the Expedition diary extracts. The!Ae!Hai Kalahari Heritage embodies this idea fully by making a spectacle of ‡Khomani San and Mier cultures, which are then viewed by tourists keen to experience the 'otherness' of their practices. As Chang and Holt argue in relation to tourism in Taiwan, the vision packaged and sold to consumers forces their 'culture into the straightjacket of a defined 'past'[and] serves to fix cultural Others in a 'timeless present', confining them to a place that cannot change, or cannot change as easily as the represer's culture' (Chang and Holt 1991, 116). In the context of peace parks, this 'Orientalising' (Said 1978) requires that only those who fit the image of indigeneity can exist in the spectacle. The majority, marginalised and impoverished, are to be excluded (see below) as these detract from the romanticised vision of the past. Put differently, the inauthenticity of the modern locates authenticity in the past, in different cultures and contexts.

For example, the development of infrastructure other than that required by the ecotourism industry is undesirable, as it can tarnish the spectacle and remind consumers that these communities are in fact present and active subjects of modernity instead of objects of the past. Representations of their culture, those that appeal to visitor tastes, are also commodified and frozen further in the production of 'African crafts' sold as memories to the parks visitors. However, these become more than 'harmless attachments to the souvenirs of destroyed cultures and dead epochs. They are also components of the conquering spirit of modernity' (MacCannell 1976, cited in Chang and Holt 1991, 115). These processes of inclusion and exclusion add a nuance to the distinction recognised by Gordon of the standard view that 'things white represent universality and things black are locked in the web of particularity' (Gordon 2006, 8). This distinction is also evident in the way that (white) western dominated environmental governance and economic institutions define the priorities and strategies for biodiversity conservation in developing countries. The majority impoverished black populations of these countries must then be able to draw on some form of 'symbolic capital', such as indigeneity, which is in itself no guarantee of inclusion, or else be subject to further alienation from the conservation-development strategies and the benefits they hold.

Accompanying the spectacle of wildlife, landscape and people in the PPF and BSA narratives is a host of 'development terminology, such as 'sustainability', 'community – empowerment', 'upliftment', and 'stakeholder participation'. These terms are the currency of donor funding and appeal to consumers of the vision that believes the parks are delivering on collaborative poverty alleviation and sustainable development. Yet, they can prove as exclusionary as they sound inclusionary precisely because they are part of a highly prescriptive conservation – development discourse, which is itself dominated by external actors who attempt to impose models from above. Büscher and Dietz (2005) argue that it is this continuity with colonial and apartheid practices that can prove more significant in the alienation of communities from conservation developments in Africa. Furthermore, the wooliness of the terms allows for fluid interpretation, giving rise to a frequent disconnect between the perceived meaning and lived experience (Wolmer 2003). For example, 'stakeholder participation' is a key tenet of the parks' JMBs strategy, but who is selected, how and on what basis is not always clear. Experience suggests that it is invariably those most willing to align with the PPF or national parks agenda, and/or those who represent the interests of selected community heads as opposed to the community as a whole; with the assumption, of course, that the community is homogenous. Moreover, and as indicated above, given that 'most marginalised and impoverished people displaced as a result of protected areas' are not indigenous (Brockington, Duffy and Igoe 2008, 123), it can be difficult for them to utilise the 'symbolic capital' (121) of indigeneity as leverage in negotiations with powerful actors to access either land, or often more importantly, other natural resources upon which their livelihoods depend.

Furthermore, what it means to participate and at what level of decision-making can also be poorly defined, subject to different power relations and also contestations within the communities apparently represented in the processes (See, for example, Chirikure et al. 2010).[20] This can result in communities feeling alienated from the process; reduced to passive observers rather than active stakeholders, which can be further exacerbated by differences in linguistic and cultural understandings, unfamiliar technocratic jargon that is hard to translate, and a disconnect in the approach to decision-making of the state and park representatives vis-à-vis the communities. It is not to suggest that efforts have not been made to address these issues but they remain, in many cases, exclusionary by practice if not intention. Research suggests also that the promise of development and poverty alleviation proffered by the vision through employment opportunities and local investment can convince communities to 'buy into' the vision through the leasing of land to state and private enterprises for conservation. Yet, the once celebrated case of the Makuleke conservancy at the heart of the Greater Limpopo Transfrontier Park[21] is testimony to how communities can be excluded from their traditional lands and resources without receiving the host of anticipated socioeconomic benefits promised by ecotourism ventures (de Villiers 1999).

Thus, the rhetoric of (eco) tourism as a mechanism for poverty alleviation in the parks is flawed.[22] Roe and Urquhart (cited in Fürsich and Robins 2004, 137) argue that 'because tourism is often driven by foreign, private sector interests, it has limited potential to contribute much to poverty elimination'. Moreover, 'most of the available jobs are unskilled, part time, casual, poorly paid, and involve long or unsociable working hours' (Craik 1997, 133). The investment of the PPF and donors in tourism sector skills training holds the promise of economic upliftment but the reality, as the Makuleke community experienced, is that employment opportunities are actually limited and job security rare (de Villiers 1999). Arguably, this could be related to the fickleness of consumerism, or the adverse impact of global economic downturns that reduce people's capacity for luxury expenditure. More significantly, however, it is because the vision is only affordable by a very small percentage of potential travellers. These complex realities detract from the idealised vision of the PPF and BSA and are thus largely absent from their narratives.

The leasing of community lands or the euphemistic 'relocation' of people from within the parks' boundaries is necessary for the spectacle of a boundless, unspoilt wilderness on the scale required by the vision. Decisions relating to who is removed, or relocated, what resources can and cannot be accessed and to what extent, are increasingly determined by the 'evidence' provided by sophisticated techno-scientific data and visual representation gathering techniques, such as Global Information Systems (GIS) which 'map' the landscape. The data is then used to support the zoning of different land-uses, the positioning of park fences, the spread of wildlife corridors and dispersal zones and so forth. The demarcation of land-use zones and park boundaries would, it is argued, be 'extremely difficult if not impossible' without them (Beech 2001). Furthermore, Craig Beech, GIS Manager at the PPF, argues that GIS provide the 'language' through which cross data-set analysis can create 'visual-spatial benchmarks' with which to measure progress and 'illustrate the significant role Peace Parks can play in lessening the impact man has had on the environment' (Beech 2006). Conveniently, what Beech does not say is which 'man' [sic]; as the tourists who fly into the parks and drive in their 4 × 4s are not considered destructive of the environment that they have come to 'save' through their consumption of it.

The visual material such technology creates also enables the PPF to market the 'concept of regional collaboration through transfrontier conservation' (Beech 2006). The production and refinement of geographical information from spatial modelling underpins the development of 'an adaptive management practice' implemented by the 'land-management authorities' (Beech 2006). Scientifically accurate 'predictive models' can then be developed in response to changing environmental conditions providing 'pertinent information and scenarios to decision makers' (Beech 2006). All of this, Beech claims, will 'help to facilitate the Peace Parks concept, and to ensure that it is accepted as a life-changing, sustainable land-use option for local communities' (Beech 2006). Again, he does qualify 'who' it is that will accept it.

Brosius and Russell (2003, 49) argue that the 'relationship between the technical and the political' is essential to understanding patterns of inclusion and exclusion, particularly as the 'evidence' produced helps to ascribe roles to actors and 'victims' of conservation (and development) agendas. While GIS make peace parks more 'legible', they also 'distance 'bioplanners' from the effects of their interventions', whilst constructing the threat to biodiversity and those posing it in a particular way (Brosius and Russell 2003, 48–49). The reductionist interpretation of the complex nexus in human–wildlife relations and associations between culture and landscape helps to delegitimise non-conservation land uses. This reinforces the colonial logic that local communities are indifferent to conservation and value the land only as an exploitable resource. The authority of such information in turn legitimates attempts to 'relocate', sometimes violently, communities whose settlements and activities are constructed as threatening to the manifestation of the vision. In the context of the Greater Limpopo TFCA, for example, the need to find dispersal spaces for the Kruger National Park's over-grown elephant population led to the exclusion of local communities from the parks through 'relocation' – not all of which has been amicable (Millgroom and Spierenburg 2008).[23] The recent launch of the PPF Climate Change Programme has further enhanced the justification – based on 'risk assessments' and 'mitigation feasibility studies' – to inhibit agricultural land-use. Instead, internationally privileged 'conservation agriculture' is favoured within and around the parks to help reduce greenhouse gas emissions (PPF *Climate Change* n.d.). Whilst reinforcing the belief that customary agricultural practices are unsustainable, these decisions do little to address the historical and contemporary experiences of structural poverty, which may (or may not) render them so. The 'evidence' however is combined with and supports 'rational' arguments about human and animal security.

This is apparent in the PPFs provision of 'electric fence worth R 250, 000 to the Maramani community of Zimbabwe to help deter stray elephants from destroying crops in the Shashe irrigation scheme' (PPF *Greater Mapungubwe* n.d.). This apparent protection of livelihoods, however, is also the 'first step in the *proper* zoning and planning of the area that will encourage the reduction of dryland cropping in sensitive wildlife dispersal areas' (PPF *Greater Mapungubwe* n.d.). This raises the question as to the underlying motive of the initiative, and whether or not it is actually a way of demarcating not where elephants may tread, but where and how the community may farm and access natural resources. This may be to the benefit of both the communities concerned and the biodiversity being conserved. However, the history of 'relocations' in the creation of the parks and the restrictions on resource use suggest rather that they can alienate people further from their traditional means of subsistence. This renders people increasingly dependent on wage labour which is neither guaranteed by the parks, nor readily available in many of the region's states. Yet such consequences can appear logical and necessary, even desirable, and thus tacitly accepted as requisite for the vision's success. Furthermore, they can reinforce historical constructions of rural African communities as poor stewards of the land.

Even the restriction of access to resources is arguably a form of displacement (Cernea 2006, 9)[24] and echoes the historical exclusion of indigenous communities from their lands in pursuit of a more 'profitable' form of land-use by the state and powerful private actors. When communities have won land claim rights their 'willingness to join the wildlife business' (PPF *Lumombo* n.d.) – to reiterate the PPF comment – can be interpreted as a processes whereby the 'governed' engage in self – improvement as defined by prevailing understandings of modernity. In doing so, they 'voluntarily pursue the goals outlined by the hegemony by fabricating the desires and preferences of "the governed", and even constructing their own subjectivity' (Tosa 2009, 427).[25] In some respects, communities such as the Nama in the Ai/Ais Richtersveld, permitted to stay because they seek to maintain their 'traditional' lifestyle, have embraced the vision in order to retain their identity and way of life. However, they are also confined by the vision in terms of how they may 'develop' and have themselves become part of the spectacle of wild Africa; considered as part of and not 'masters' of the land.[26]

Moreover, and like the development-speak evident in the BSA and PPF literature discussed above, the language provided by GIS is common only to a select group of elite actors. It is not the language of the communities living within or adjacent to the parks, yet it is privileged over other forms of mapping, such as the counter-mapping of indigenous knowledge undertaken by the people themselves (see for example The African Biodiversity Network and Gaia Foundation, 2011). Such techniques can provide a more in-depth and sociologically rich understanding of human interactions with the landscape and wildlife of the parks that might better inform, or at the very least compliment, the highly abstracted information produced by GIS. They also have the potential to genuinely empower communities as active participants in decision-making processes by respecting their agency and providing the platform for the expression of their knowledge and experience about local ecology, as well as their own cultural and livelihood needs.

At the outset of this paper, it was suggested that the borderzones of Southern Africa were highly contested and historically violent spaces on the periphery of the state, but that peace parks reconceptualised them as spaces of opportunity. What the peace parks vision has done is define the nature and physicality of that opportunity. Moreover, adapting the argument made by Peluso (1993, 199), they have allowed the state and private actors to hijack the 'ideology, legitimacy and technology' of conservation to increase or appropriate 'their control over valuable resources and recalcitrant populations'. For example, whilst the construction of fences to 'protect' tourists and animals and entry gates through which park visitors and workers pass symbolise access and the 'openness' of the spaces beyond, they are also clear makers of state (via their national parks authorities) and, it could be argued, PPF authority. Moving rather than removing the fences, and sometimes by substantial distances, also requires a greater policing presence to cover the range of potential threats – from illegal travellers, poachers and illicit traders – and more fundamentally anyone else that is not a legitimate consumer of the parks.

In Southern Africa, the reality is thus not one of a 'boundless' landscape but the fencing in of vast conservation spaces and inversely, the fencing out of particular groups of people. What determines admittance is not ones' citizenship of the participating countries, but the capacity to pay and the means to travel. However, even relatively wealthy people may be excluded from enjoying the transfrontier experience as many crossing routes require 4 × 4 capabilities. Visitors must also pass through police checkpoints and border patrols, which are often located in the less accessible parts of the TFCAs. For example, the border post connecting South Africa and Mozambique in the Greater Limpopo Park is located in the less visited northern section of the Kruger National Park and requires 4 × 4 transport capabilities. The border posts are not only a reminder of the state's presence in the parks, but act as controls on the freedom of people within them. Thus, even those privileged enough to cross the border are subjected to a series of identity checks, including the recording of identification numbers, copying of passports and travel plans and tracking of vehicle registrations; all of which are designed to monitor those within the parks and exclude unwanted elements, but are routinised and accepted as both normal and necessary.

Furthermore, in the Ai/Ais Richtersveld TFCA, the PPF has sought to 'better control access from the south to the Namibian section of the Transfrontier Park' through the construction of an 'access control facility' at Gamkop (PPF *Ai/Ais*/n.d.). In this case, it is the PPF that is able to dictate entry in and out of the park, giving it the power to define who constitutes a legitimate visitor based on prescribed understandings. Another mechanism for inhibiting cross border transience has been the stipulation, in the Kgalagadi for example, that to cross the border requires the booking of one night's accommodation in the park to reduce commercial traffic along the new road linking the park sections (PPF *Kgalagadi* n.d.). Such traffic would, of course, not fit with the vision of an untouched and relatively human free wilderness as described by the PPF (above). Selective access is therefore, determined both by the physical policing of the space and by the exclusivity of the commodity being consumed.

Ellis (1994) claims that an 'element of coercion' is necessary in the construction of national, and therefore by extension, international parks. A process he argues, which has continued in the post-apartheid era with South African Defence Forces'

(SADF) training of game wardens in the region and the *koevoet* counter-insurgency troops deployed for conservation purposes in Namibia in the 1990s (67).[27] Furthermore, the creation of the parks and the 'security' of both visitors and animals – particularly endangered animals – have increased border patrols and the establishment of anti-poaching units (Spenceley and Schoon 2007). As conservation initiatives in the region have historically involved, rather controversially, the use of military and or police personnel, it would be unsurprising if local communities associate peace parks with the violent history of police and military brutality and restricted access to their natural resource base. This is especially the case when people traditionally dependent on hunting as part of their cultural and livelihood practices are now conflated with those exploiting the biodiversity to fuel the illicit international trade of ivory and other rare flora and fauna. This is particularly challenging when the hunting communities pursue species now endangered as a result of the illicit trade. Thus, local communities are constructed as either 'conservation heroes' or 'environmental villains' (Moore 2010, 19) primarily because of (internationally developed) policy decisions, underpinned by scientific 'evidence' about conservation priorities that discredit indigenous cultural practices and local community knowledge about natural resource management. Except, that is, when it can be incorporated into tourism activities.

Conclusion

In conclusion, the construction and articulation of the peace parks vision draws legitimacy from a play on fear about environmental degradation, and hopes about biodiversity conservation and the future security and prosperity of Southern Africa's states and their inhabitants. The representation and performativity of peace parks is, in turn, informed by dominance of neoliberalism and in particular the commodification of the 'spectacle' of nature and culture that has become integral to prevailing environmental governance and conservation strategies. However, the vision of Southern African peace parks is only a partial albeit powerful representation of reality that operates like the 'exception' to discipline the region in line with the vision and the interests it serves. The result is the blending of the popular and alluring (mis)conception of 'wild' Africa and exoticism attached to 'traditional' communities with the trappings and comforts of modern-day travel and adventure.[28] However, when market-based solutions are presented – globally they increasingly are – as the most logical and effective means through which to achieve a host of environmental conservation and development needs, the actual and complex 'reality' of the lived life of the parks can be easily obscured (McAfee 1999).[29] Without effective 'management' of the lived life of the parks, the vision and all that it promises is continually threatened. Thus, to borrow from the words of Tosa (2009, 414) 'neoliberal governmentality promotes securitisation of supposedly risky groups on the periphery' and the adoption of illiberal practices in the pursuit of security of the referent; in this case, the vision of the peace parks.

As such, the commodification of nature and culture in Southern Africa's peace parks creates patterns of inclusion and exclusion, security and insecurity, which reinforce rather that break with historical processes of alienation – physical, economic, political and psychological – of marginalised people at the periphery of the state. Moreover, while there appears to be a move towards 'engaging' local communities in the development and management of Southern African peace parks – as evidenced by the PPF and BSA materials – it is difficult not to see how these claims of inclusivity are actually also disciplining the people and landscape in line with the market. They are only inclusive if people conform to the 'vision' for which their participation is desired. Thus, processes of inclusion and exclusion have become naturalised in the daily management of the parks and legitimated by the overarching international and regional objectives for which the vision was constructed. However, dependency on a volatile and unpredictable commodity driven economy that feeds off the 'otherness' of people and place, also exposes the region and its conservation and development aspirations to the fickleness of modern consumerism. Thus, claims by the PPF and region's states (through BSA) to overcome a multiplicity of historical injustices ring hollow, when the very approaches adopted to do so are grounded in a global economic order dominated by external actors who continue to view Africa and its people, as somehow outside of modernity.

Acknowledgements

The author would like to thank Prof. Bram Büscher for his critical and helpful comments in the development of this paper, the research assistance of Sarah Bruchhausen and the insightful comments of the JCAS reviewers.

Notes

1. The 'politics of exception' refers to discourses of insecurity and extraordinary political measures adopted or curtailed because of the invocation of the 'emergency'. Gorgio Agamben's (1998) writing on the 'state of exception' has been the foundation for the theorising of the exception in poststructuralist security studies which have informed the interpretation of the concept in this paper. See, for example Aradau (2004), Huysmans and Buonfino (2008) and specifically on the role of borders and spaces of exclusion, Basaran (2008) and Vaughan-Williams (2008, 2009).
2. Benedict Anderson ([1983] 1991) provides a thought-provoking discussion on the socially constructed notion of a nation as a limited but sovereign form of community imagined by its members.
3. The term apartheid is associated specifically with South Africa, but similar experiences were experienced in other countries in the region, most notably Namibia and Zimbabwe.
4. See Brockington, Duffy, and Igoe (2008) for an enlightening use of Guy Debord's Society of the Spectacle ([1967] 1995) to explain how nature-as-spectacle is commodified. Igoe (2010, 375) argues that the commodification of the spectacle alienates people from the 'processes that produced them' and that consumers are equally ignorant of these processes.
5. In a paper presented at the Association of American Geographers Annual Conference in Seattle, 2011 the author argued that it is the economic imperative that is ultimately privileged over the conservation agenda precisely because of the powerful framing narrative of the market in the development of Southern African peace parks (Barrett 2011).
6. See also Perez et al. (2007) and Jindal, Swallow, and Kerr (2008).
7. Samir Amin (2011) argues that maldevelopment is a more appropriate term than underdevelopment, implying that there was, and continues to be, a deliberate process of exploitation by the majority world states and institutions of the minority world and particularly former colonies.
8. For a critical discussion of this see van Amerom and Büscher (2005).
9. Ali's (2007) edited volume discusses the potential advantages of transboundary conservation initatives in greater detail.

10. Spierenburg and Wels (2010) explore the 'social capital' of the late Anton Rupert and the late Prince Bernhard of the Netherlands in attracting funding and support for transfrontier conservation in Southern Africa and how influential they have been in framing conservation strategies. The PPF has continued to attract substantial funding from European state bodies and from the South African government. See PPF (*Donors of Peace Parks Foundation* n.d.) for more information, as well as the organisation's annual reports and financial statements (2006, 2009, *Annual Reviews* n.d.).

11. The PPF also gained 'NGO Observer Status' at the United Nations Climate Change Conference in Cancun in December 2010 (PPF *Climate Change* n.d.).

12. The term 'majority world' refers to the majority of humanity, primarily but not exclusively in former colonies, who are economically and politically marginalised. In contrast, the term 'minority world' refers to the economic and political elite of humanity primarily but not exclusively located in the more advanced industrial states of Europe, North America, Australia and Japan. Although not intended to obscure the many variances of experience covered by such labels, the term is arguably more useful and reflective of global inequalities and power relations than the more commonly used 'Global North' and 'Global South', 'developed' and 'underdeveloped', or 'first' and 'third' worlds. These latter definitions reinforce the global structures of power and situate people and states in a hierarchy that has historically been ideologically and materially constructed by the minority world states of Europe and North America. See, for example, John Morgan (2001, 10).

13. A comment which appears to contradict the fact that San communities have lived in the Kalahari for centuries.

14. Thabo Mbeki (2004) first presented his idea of an African Renaissance in his now infamous 'I am an African' speech in which he sought to challenge regressive constructions of Africa and Africans by invoking visions of a culturally and intellectually rich pre-colonial continent and the potential within Africa to deliver its own development.

15. The BSA lyrics are particularly telling: Africa! Africa! Southern Africa! Oceans, mountains, deserts, and plains/Wherever we go/Boundless and free/Transfrontier, together we work/Tirelessly we keep and conserve/The wonder we have, for them to come/ Boundless they roam, protected and free (BSA *Boundless Song* n.d.). The Boundless song is reflective of the selective use of African oral traditions in the construction of a spectacle for a commercial end.

16. Kevin Dunn's (2009) work is particularly insightful. He illustrates how state structuring/ structural effects are discursively produced through continual acts of performativity. His article not only interrogates these processes of performativity in the context of African national parks but argues that the continual need for them creates opportunities for resistence and contestation of state power and authority.

17. Poverty is widely critiqued as a major threat to environmental integrity. See Duraiappah (1996) for a review of the debate.

18. This is an unfortunate acronym given that communities have been 'relocated' because of the parks.

19. Governmentality is employed here in the Foucauldian sense; as the different mentalities, rationalities and techniques used to govern people which can be understood as a collective process of disciplining.

20. In relation to the Makuleke, see: de Villiers (1999) and Steenkamp and Uhr (2000).

21. The author recently visited the Kruger National Park and entered the Makuleke conservancy area now leased to Wilderness Safaris. A sign near the Wilderness camp read 'Makuleke Contractual Land; Heart of the Trasnsfrontier Park'.

22. This is not to suggest that all pro-poor initiatives are flawed although there are key examples from the region – such as CAMPFIRE – in Zimbabwe which was an early attempt to engage local communities in wildlife management for livelihood development but which, subsequently, became the target of criticism for being unsuccessful (See, for example Alexander and McGregor 2000).

23. In a critical analysis of the 'relocation' of 26,000 Mozambican from the land designated for the Limpopo National Park, later incorporated into the Greater Limpopo Transfrontier Conservation Area, Millgroom and Spierenburg (2008) argue that the

restrictions on livelihood strategies resulted in an induced acceptance of relocation. Thus, the expansion of wildlife dispersal areas, park regulations and management procedures proved effective at excluding local communities from these spaces. Moore (2010, 19) also argues that environmental narratives are being used as 'tools of persuasion' to garner African people's compliance with international environmental agendas.

24. Curran et al. (2009) have contended that in the last decade there has been no 'forced' relocation of peoples in the creation of national parks in Africa but the issue remains contentious and as Cernea (2006) highlights, displacement may not mean the physical exclusion from the land but denial of access to resources or imposed regulations about land-use options which inhibit 'traditional' livelihoods and result in 'relocation'.

25. Tosa (2009) discusses this idea of 'voluntary' subjugation in his analysis of the 'global slum'.

26. See, for example, Patrick Harries' (2007) excellent analysis of the Swiss Missionary construction of the 'African' and 'Africa' and the relationship between European and African knowledge systems.

27. Koevoet was a group of primarily Zulu personnel set up by the apartheid security police as the main counter-insurgency unit in Namibia prior to its independence.

28. Although not discussed in detail here the parks' amenities are intended to be 'modern' without losing their 'traditional appeal'.

29. McAfee (1999) argues that this is a common trend in the World Bank and was evident in the development of the Convention on Biodiversity Conservation. Spatial and social contexts are obscured because governance processes locate conservation 'best- practice' within international (consumer) markets.

Note on contributor

George Barrett is a Lecturer and PhD candidate in the Department of Political and International Studies at Rhodes University in South Africa. Her main research interests include critical security studies, environmental politics and ethics, and an international political sociology of security with specific attention placed on the way in which neoliberal market capitalism and the commodification of life are creating and recreating processes of inclusion and exclusion, security and insecurity. George is also interested in processes of political change and to this end teaches a second year undergraduate course called Revolutions in Comparative Perspective. Her other teaching currently includes a third year undergraduate course on Environmental Politics and Ethics and a Masters/Hons course called Liberation Ecologies although she has also taught on International Relations, Global Environmental Governance and the United Nations system. George is also part of the Thinking Africa research and teaching programme, which seeks to promote African theory and scholarship both within the continent and also on the international stage.

References

African Wildlife Foundation (AWF). 2002. "Great Limpopo Transfrontier Park will Double Land for Wildlife- Including Kruger's 10,000 Elephants." African Wildlife Foundation. Accessed April 29, 2011. http://www.awf.org/content/headline/detail/1163.html

Agamben, Giorgio. 1998. *Homo Sacer: Sovereign Power and Bare Life*. Stanford: Stanford University Press.

Alexander, Jocelyn, and JoAnn McGregor. 2000. "Wildlife and Politics: CAMPFIRE in Zimbabwe." *Development and Change* 31: 605–627. doi:10.1111/1467-7660.00169.

Ali, Saleem H., ed. 2007. *Peace Parks: Conservation and Conflict Resolution*. Cambridge, MA: The MIT Press.

Amin, Samir. 2011. *Maldevelopment: Anatomy of a Global Failure*. 2nd ed. Cape Town: Pambazuka Press.

Anderson, Benedict. [1983] 1991. *Imagined Communities: Reflections on the Origin and Spread of Nationalism*. London: Verso.

Aradau, Claudia. 2004. "Security and the Democratic Scene: Desecuritization and Emancipation." *Journal of International Relations and Development* 7 (4): 388–413. doi:10.1057/palgrave.jird.1800030.

Barrett, George. 2011. "Utopian Dreams, Apocalyptic Nightmares and Market Panaceas." Paper presented at the Association of American Geographers Annual Conference, Seattle, April 12.

Basaran, Tugba. 2008. "Security, Law, Borders: Spaces of Exclusion." *International Political Sociology* 2 (4): 339–354. doi:10.1111/j.1749-5687.2008.00055.x.

Beech, Craig. 2001. "ESRI- ArcNews". The Peace Parks Foundation Aided by GIS. Accessed May 7, 2011. http://www.esri.com/news/arcnews/summer01articles/thepeaceparks.html.

Beech, Craig. 2006. "Peace Parks Foundation GIS Lecture, Oxford University." African Environments Programme, University of Oxford. Accessed May 1, 2011. http://www.african-environments.ouce.ox.ac.uk/events/2006/190906.doc.

Bigo, Didier. 2008. "International Political Sociology." In *Security Studies: An Introduction*, edited by Paul D. Williams, 116–129. London: Routledge.

Blandy, Fran. 2006. "Transfrontier Parks a Boost to Conservation." IOL SciTech. Accessed February 21, 2011. http://www.iol.co.za/scitech/technology/transfrontier-parks-a-boost-to-conservation-1.282687

BSA (Boundless Southern Africa). n.d. "/Ai/Ais- Richtersveld Transfrontier Conservation Area." Boundless Southern Africa. Accessed April 29, 2011. http://www.boundlessa.com/en/index.php?option=com_content&task=view&id=141&Itemid=257.

Brockington, Dan, Rosaleen Duffy, and James Igoe. 2008. *Nature Unbound; Conservation, Capitalism and the Future of Protected Areas.* London and Washington, DC: Earthscan.

Brosius, J. Peter, and Diane Russell. 2003. "Conservation from Above: An Anthropological Perspective on Transboundary Protected Areas and Ecoregional Planning." *Journal of Sustainable Forestry* 17 (1–2): 39–65. doi:10.1300/J091v17n01_04.

BSA (Boundless Southern Africa). 2008. *Boundless 2008.* Boundless Southern Africa. Accessed April 27, 2011. http://www.boundlessa.com/en/index.php?option=com_content&task=view&id=172&Itemid=308

BSA (Boundless Southern Africa). 2009. *Boundless Southern Africa Expedition 2009*. Accessed May 13, 2011. http://www.boundlessa.com/en/index.php?option=com_content&task=blogcategory&id=17&Itemid=306.

BSA (Boundless Southern Africa). n.d. *Boundless Song*. Boundless Southern Africa. Accessed April 29, 2011. http://www.boundlessa.com/en/index.php?option=com_content&task=view&id=172&Itemid=308.

BSA (Boundless Southern Africa). n.d. *Kavango-Zambezi Transfrontier Conservation Area*. Boundless Southern Africa. Accessed April 29, 2011. http://www.boundlessa.com/en/index.php?option=com_content&task=view&id=139&Itemid=253.

BSA (Boundless Southern Africa). n.d. *Kgalagadi Transfrontier Park*. Boundless Southern Africa. Accessed April 29, 2011. http://www.boundlessa.com/en/index.php?option=com_content&task=view&id=138&Itemid=250.

Büscher, Bram. 2010a. "Derivative Nature: Interrogating the Value of Conservation in 'Boundless Southern Africa.'" *Third World Quarterly* 31 (2): 259–276. doi:10.1080/01436591003711983.

Büscher, Bram. 2010b. "Seeking 'Telos' in the 'Transfrontier'? Neoliberalism and the Transcending of Community Conservation in Southern Africa." *Environment and Planning A* 42 (3): 644–660. doi:10.1068/a42140.

Büscher, Bram, and Ton Dietz. 2005. "Conjunctions of Governance: The State and the Conservation-Development Nexus in Southern Africa." *The Journal of Transdisciplinary Environmental Studies* 4 (2): 1–15. http://repub.eur.nl/res/pub/32289/JTES200541.pdf.

Büscher, Bram, and Wolfram Dressler. 2007. "Linking Neoprotectionism and Environmental Governance: On the Rapidly Increasing Tensions between Actors in the Environment-Development Nexus." *Conservation and Society* 5 (4): 586–611. http://www.conservationandsociety.org/temp/ConservatSoc54586-4253782_114857.pdf.

Büscher, Bram, and Wolfram Dressler. 2012. "Commodity Conservation. The Restructuring of Community Conservation in South Africa and the Philippines." *Geoforum* 43 (3): 367–376. doi:10.1016/j.geoforum.2010.06.010.

Carrier, James G. 2010. "Protecting the Environment the Natural Way: Ethical Consumption and Commodity Fetishism." *Antipode* 42 (3): 672–689. doi:10.1111/j.1467-8330.2010. 00768.x.

Cernea, Michael. M. 2006. "Population Displacement Inside Protected Areas: A Redefinition of Concepts in Conservation Policies." *Policy Matters* 14: 8–26.

Chang, Hui-Ching, and G. Richard Holt. 1991. "Tourism as Consciousness of Struggle: Cultural Represents of Taiwan." *In Critical Studies in Mass Communication* 8 (1): 102–118. doi:10.1080/15295039109366783.

Chirikure, Shadreck, Munyaradzi Manyanga, Webber Ndoro, and Gilbert Pwiti. 2010. "Unfulfilled Promises? Heritage Management and Community Participation at some of Africa's Cultural Heritage Sites." *International Journal of Heritage Studies* 16 (1–2): 30–44. doi:10.1080/13527250903441739.

Craik, Jennifer. 1997. "The Culture of Tourism." In *Touring Cultures: Transformation of Travel and Theory*, edited by Chris Rojek and John Urry, 133. London: Routledge.

Cronon, William. 1996. "The Trouble with Wilderness: Or, Getting Back to the Wrong Nature." *Environmental History* 1 (1): 7–28. doi:10.2307/3985059.

Curran, Bryan, Terry Sunderland, Fiona Maisels, John Oates, Stella Asaha, Michael Balinga, Louis Defo, et al. 2009. "Are Central Africa's Protected Areas Displacing Hundreds of Thousands of Rural Poor?" *Conservation and Society* 7: 30–45. doi:10.4103/0972-4923. 54795.

Debord, Guy. [1967] 1995. *Society of the Spectacle*. New York: Zone Books.

de Villiers, Bertus. 1999. *Land Claims & National Parks: The Makuleke Experience*. Pretoria: Human Sciences Research Council.

Dressler, Wolfram, and Bram Büscher. 2008. "Market Triumphalism and the CBNRM 'Crises' at the South African Section of the Great Limpopo Transfrontier Park." *Geoforum* 39 (1): 452–465. doi:10.1016/j.geoforum.2007.09.005.

Duffy, Rosaleen. 2005. *Global Politics and Peace Parks. Parks for Peace or Peace for Parks*. Washington, DC: Woodrow Wilson International Center for Scholars: 6.

Duffy, Rosaleen. 2007. "Peace Parks and Global Politics: The Paradoxes and Challenges of Global Governance." In *Peace Parks: Conservation and Conflict Resolution*, edited by Saleem H. Ali, 55–68. London: MIT Press.

Dunn, Kevin C. 2009. "Contested State Spaces: African National Parks and the State." *European Journal of International Relations* 15 (3): 423–446. doi:10.1177/ 1354066109338233.

Duraiappah, Anantha K. 1996. *CREED Working Paper Series No 8: Poverty and Environmental Degradation: A Literature Review and Analysis*. London: International Institute for Environment and Development.

Ellis, Stephen. 1994. "Of Elephants and Men: Politics and Nature Conservation in South Africa." *Journal of Southern African Studies* 20 (1): 53–69. doi:10.1080/03057079408708386.

Fox, Justin. 2009. "Kaza, the Birth of Africa's Greatest Game Park." *Getaway* 34–43. http:// magazine.getaway.co.za/archive/kaza-birth-africas-greatest-game-park/.

Fürsich, Elfriede, and Melinda B. Robins. 2004. "Visiting Africa: Constructions of Nation and Identity on Travel Websites." *In Journal of Asian and African Studies* 39: 133–152. doi:10.1177/0021909604048255.

Gertruida, Tannie. 2009. "#19– Tannie Gertruida and the Nama Stap." *Boundless Southern Africa, Expedition News*. Accessed April 29, 2011. http://www.boundlessa.com/en/index. php?option=com_content&task=view&id=195&Itemid=306.

Gordon, Lewis R. 2006. "African American Philosophy, Race and the Geography of Reason." In *Not Only the Master's Tools*, edited by Lewis R. Gordon and Jane A. Gordon, 3–50. Colorado: Paradigm.

Hanks, John. 2003. "Transfrontier Conservation Areas (TFCAs) in Southern Africa: Their Role in Conserving Biodiversity, Socioeconomic Development and Promoting a Culture of Peace." *Journal of Sustainable Forestry* 17 (1–2): 127–148. doi:10.1300/J091v17n01_08.

Harries, Patrick. 2007. *Butterflies and Barbarians: Swiss Missionaries and Systems of Knowledge in South-East Africa*. Oxford: James Curry Publishers.

Holgate, Kingsley. 2011. "The Grey Beard of Africa." *Kingsley Holgate*. Accessed April 27, 2011. http://www.kingsleyholgate.net/about-kingsley-holgate.html.

Hughes, David McD. 2002. "Going Transboundary: Scale-Making and Exclusion in Southern-African Conservation." Paper Presented at the Environment and Development Advanced Research Circle, University of Wisconsin, Madison, USA. Accessed February 13, 2013. http://hdgc.epp.cmu.edu/misc/Going%20Transboundary%20-%20Hughes%5B1%5D.pdf.

Huysmans, Jef, and Alesandra B. Buonfino. 2008. "Politics of Exception and Unease: Immigration, Asylum and Terrorism in Parliamentary Debates in the UK." *Political Studies* 56 (4): 766–788. doi:10.1111/j.1467-9248.2008.00721.x.

Igoe, James. 2010. "The Spectacle of Nature in the Global Economy of Appearances: Anthropological Engagements with the Spectacular Mediations of Transnational Conservation." *Critique of Anthropology* 30 (4): 375–397. doi:10.1177/0308275X10372468.

Igoe, James, and Dan Brockington. 2007. "Neoliberal Conservation: A Brief Introduction." *Conservation and Society* 5 (4): 432–449. http://www.conservationandsociety.org/temp/ConservatSoc54432-4368659_120806.pdf.

Igoe, James, Katja Neves, and Dan Brockington. 2010. "A Spectacular Eco-Tour around the Historic Bloc: Theorising the Convergence of Biodiversity Conservation and Capitalist Expansion." *Antipode* 42 (3): 486–512. doi:10.1111/j.1467-8330.2010.00761.x.

Jindal, Rohit, Brent Swallow, and John Kerr. 2008. "Forestry-Based Carbon Sequestration Projects in Africa: Potential Benefits and Challenges." *Natural Resources Forum* 32 (2): 116–130. doi:10.1111/j.1477-8947.2008.00176.x.

King, David A., and William P. Stewart. 1996. "Ecotourism and Commodification: Protecting People and Places." *Biodiversity Conservation* 5 (3): 293–305. doi:10.1007/BF00051775.

MacDonald, Kenneth I. 2010. "The Devil is in the (Bio) Diversity: Private Sector "Engagement" and the Restructuring of Biodiversity Conservation." *Antipode* 42 (3): 513–550. doi:10.1111/j.1467-8330.2010.00762.x.

Mbeki, Thabo. 2004. "I Am an African." *African Renaissance* 1 (2): 9–13. http://www.hollerafrica.com/pdf/vol1AfricanRenSep_Oct_2004.pdf#page=9.

McAfee, Kathleen. 1999. "Selling Nature to Save it? Biodiversity and Green Developmentalism." *Environment and Planning D: Society and Space* 17 (2): 133–154. doi:10.1068/d170133.

MCSA (Media Club South Africa). 2009. "Famous Adventurer Promotes Transfrontier Conservation." *Africa Good News.* June 18. Accessed February 17, 2011. http://www.africagoodnews.com/development/education/615-famous-adventurer-to-promote-transfrontier-conservation.html.

Millgroom, Jessica, and Marja Spierenburg. 2008. "Induced Volition: Resettlement from the Limpopo National Park, Mozambique." *Journal of Contemporary African Studies* 26 (4): 435–448. doi:10.1080/02589000802482021.

MoEAT (Ministry of Environmental Affairs and Tourism). 2001. "Minister of Environmental Affairs and Tourism, Valli Moosa, Introduces the Great Limpopo Transfrontier Park." *South African Government Information.* Accessed April 29, 2011. http://www.info.gov.za/speeches/2001/0110081046a1002.html.

Moore, Lorraine E. 2010. "Conservation Heroes Versus Environmental Villains: Perceiving Elephants in the Caprivi Strip." *Human Ecology* 38 (1): 19–29. doi:10.1007/s10745-009-9290-x.

Morgan, John. 2001. *Development, Globalisation and Sustainability.* London: Nelson Thornes.

Myburgh, Werner. n.d. "Message from the CEO of the Peace Parks Foundation." *Peace Parks Foundation.* Accessed February 21, 2011. http://www.peaceparks.org/story.php?pid=1&mid=3.

OAU (Organization of African Unity). 1981. *African Charter on Human and Peoples' Rights ("Banjul Charter")*, CAB/LEG/67/3 rev. 5, 21 I.L.M. 58 (1982). Accessed January 15, 2013. http://www.unhcr.org/refworld/docid/3ae6b3630.html.

Peluso, Nancy L. 1993. "Coercing Conservation? The Politics of State Resource Control." *Global Environmental Change* 3 (2): 199–217. doi:10.1016/0959-3780(93)90006-7.

Peoples, Columba, and Nick Vaughan-Williams. 2010. "Poststructuralism and International Political Sociology." In *Critical Security Studies: An Introduction*, edited by Columbus Peoples and Nick Vaughan-Williams, 62–74. Abingdon: Routledge.

Perez, Carlos A., Constance Neely, Carla Roncoli, and Jean Steiner, eds. 2007. Special Issue: Making Carbon Sequestration Work for Africa's Rural Poor: Opportunities and Constraints. *Agricultural Systems* 94 (1): 1–110.

PPF (Peace Parks Foundation). 2006. *Annual Review and Financial Statement*. Stellenbosch, South Africa: Peace Parks Foundation.

PPF (Peace Parks Foundation). 2009. *Annual Review and Financial Statement*. Stellenbosch, South Africa: Peace Parks Foundation.

PPF (Peace Parks Foundation). n.d. *!Ae!Hai Kalahari Heratige Park*. Peace Parks Foundation. Accessed January 15, 2013. http://www.peaceparks.org/programme.php?pid=25&mid=1112.

PPF (Peace Parks Foundation). n.d. */Ai/Ais Richtersveld Transfrontier Park*. Peace Parks Foundation. Accessed April 29, 2011. http://www.peaceparks.org/tfca.php?pid=27&mid=1001

PPF (Peace Parks Foundation). n.d. *Annual Reviews and Financial Statements*. Peace Parks Foundation. Accessed January 23, 2011. http://www.peaceparks.org/story.php?pid=1&mid=17

PPF (Peace Parks Foundation). n.d. *Climate Change Programme in Southern African TFCAs*. Peace Parks Foundation. Accessed February 12, 2011. http://www.peaceparks.org/programme.php?pid=24&mid=1021.

PPF (Peace Parks Foundation). n.d. *Donors of Peace Parks Foundation*. Peace Parks Foundation. Accessed January 23, 2011. http://www.peaceparks.org/story.php?pid=42&mid=50.

PPF (Peace Parks Foundation). n.d. *Great Limpopo Transfrontier Park*. Peace Parks Foundation. Accessed April 29, 2011. http://www.peaceparks.org/tfca.php?pid=27&mid=1005

PPF (Peace Parks Foundation). n.d. *Greater Mapungubwe*. Peace Parks Foundation. Accessed April 29, 2011. http://www.peaceparks.org/tfca.php?pid=27&mid=1003.

PPF (Peace Parks Foundation). n.d. *Kavango-Zambezi*. Peace Parks Foundation. Accessed April 27, 2011. http://www.peaceparks.org/tfca.php?pid=27&mid=1008.

PPF (Peace Parks Foundation). n.d. *Kgalagadi Transfrontier Park*. Peace Parks Foundation. Accessed April 29, 2011. http://www.peaceparks.org/tfca.php?pid=27&mid=1002

PPF (Peace Parks Foundation). n.d. *Lumombo Transfrontier Conservation and Resource Area*. Peace Parks Foundation. Accessed April 29, 2011. http://www.peaceparks.org/tfca.php?pid=27&mid=1006.

PPF (Peace Parks Foundation). n.d. *Lower Zambezi-Mana Pools TFCA*. Peace Parks Foundation. Accessed April 27, 2011. http://www.peaceparks.org/tfca.php?pid=27&mid=1019

PPF (Peace Parks Foundation). n.d. *Maloti-Drakensberg Transfrontier Conservation and Development Area*. Peace Parks Foundation. Accessed April 29, 2011. http://www.peaceparks.org/tfca.php?pid=27&mid=1004.

PPF (Peace Parks Foundation). n.d. *Origins*. Peace Parks Foundation. Accessed February 21, 2011. http://www.peaceparks.org/story.php?pid=1&mid=2.

PPF (Peace Parks Foundation). n.d. *Why Peace Parks?* Peace Parks Foundation. Accessed April 27, 2011. http://www.peaceparks.org/story.php?pid=19&mid=20

PPF (Peace Parks Foundation). n.d. *Training Colleges*. Peace Parks Foundation. Accessed January 15, 2013. http://www.peaceparks.org/story.php?pid=100&mid=28

PPF (Peace Parks Foundation). n.d. *Southern African Peace Parks? The TFCAs*. Peace Parks Foundation. Accessed April 29, 2011. http://www.peaceparks.org/story.php?pid=100&mid=19

Rumble, Mike. n.d. "Message from Mike Rumble." *Boundless Southern Africa Expedition*. Accessed May 11, 2011. http://www.imagineering.co.za/boundlesssa/mike.html.

SADC (Southern African Development Community). n.d. *Transfrontier Conservation Areas (TFCAs)*. Accessed May 12, 2011. http://www.sadc.int/fanr/naturalresources/transfrontier/index.php?media=print&media=print.

Sandilands, Catriona. 1993. "On 'Green' Consumerism: Environmental Privitization and 'Family Values.'" *Canadian Women's Studies/les cahiers de la femmes* 13 (3): 45–47. http://pi.library.yorku.ca/ojs/index.php/cws/article/download/10409/9498.

147

Sandwith, Trevor, Clare Shine, Lawrence Hamilton, and David Sheppard. 2001. *Transboundary Protected Areas for Peace and Co-operation*. (7) IUCN-World Conservation Union.

Said, Eward. 1978. *Orientalism*. London: Penguin.

SANParks (South African National Parks). n.d. "Great Limpopo Transfrontier Park." *Conservation Services*. Accessed January 15, 2013. http://www.sanparks.org/conservation/transfrontier/great_limpopo.php.

Schwartz, Mark W. 1999. "Choosing the Appropriate Scale of Reserves for Conservation." *Annual Review of Ecology and Systematics* 30 (1): 83–108. doi:10.1146/annurev.e-colsys.30.1.83.

Spenceley, Anna, and Michael Schoon. 2007. "Peace Parks as Social Ecological Systems: Testing Environmental Resilience in Southern Africa." In *Peace Parks: Conservation and Conflict Resolution*, edited by Saleem H. Ali, 83–104. Cambridge, MA: The MIT Press.

Spenceley, Anna. 2006. "Tourism in the Great Limpopo Transfrontier Park." *Development Southern Africa* 23 (5): 649–667. doi:10.1080/03768350601021897.

Spierenburg, Marja, and Harry Wels. 2010. "Conservative Philanthropists, Royalty and Business Elites in Nature Conservation in Southern Africa." *Antipode* 42 (3): 647–670. doi:10.1111/j.1467-8330.2010.00767.x.

Steenkamp, Conrad, and Jana Uhr. 2000. "The Makuleke Land Claim: Power relations and Community-Based Natural Resource Management." *Evaluating Eden Series, IIED* 18: 1–26. http://conservation-development.net/Projekte/Nachhaltigkeit/CD1/Suedafrika/Literatur/PDF/Steenkamp2000.pdf.

Stevens, P. W., and R. Jansen. 2002. *Botswana National Ecotourism Strategy*. Gaborone: The Government of Botswana.

The African Biodiversity Network and The Gaia Foundation. 2011. *Reviving Our Culture, Mapping Our Future*. New York: Vimeo. Online Video. Accessed January 15, 2013. http://www.gaiafoundation.org/galleries/videos/reviving-our-culture-mapping-our-future.

Thorsell, Jim W., ed. 1991. *Parks on the Borderline: Experiences in Transfrontier Conservation*. Gland, Switzerland: International Union for the Conservation of Nature.

Tosa, Hiroyuki. 2009. "Anarchical Governance: Neoliberal Governmentality in Resonance with the State of Exception." *International Political Sociology* 3 (4): 414–430. doi:10.1111/j.1749-5687.2009.00084.x.

Vale, Peter. 2003. *Security and Politics in Southern Africa: The Regional Dimension*. Boulder, CO: Lynne Rienner.

van Amerom, Marloes, and Bram Büscher. 2005. "Peace Parks in Southern Africa: Bringers of an African Renaissance?" *The Journal of Modern African Studies* 43 (2): 159–182. http://dx.doi.org/10.1017/S0022278X05000790.

Vaughan-Williams, Nick. 2008. "Borders, Territory, Law." *International Political Sociology* 2 (4): 322–338. doi:10.1111/j.1749-5687.2008.00054.x.

Vaughan-Williams, Nick. 2009. *Border Politics: The Limits of Sovereign Power*. Edinburgh: Edinburgh University Press.

West, Paige, James Igoe, and Dan Brockington. 2006. "Parks and Peoples: The Social Impact of Protected Areas." *Annual Review of Anthropology* 35 (1): 251–277. doi:10.1146/annurev.anthro.35.081705.123308.

Westling, Arthur H., ed. 1993. *Transfrontier Reserves for Peace and Nature: A Contribution to Human Security*. Nairobi: United Nations Environment Programme.

Wolmer, William. 2003. "Transboundary Conservation: The Politics of Ecological Integrity in the Great Limpopo Transfrontier Park." *Journal of Southern African Studies* 29 (1): 261–278. doi:10.1080/0305707032000060449.

New geographies of conservation and globalisation: the spatiality of development for conservation in the iSimangaliso Wetland Park, South Africa

Melissa Hansen[a,b]

[a]Lund University Centre of Excellence for the Integration of Social and Natural Dimensions of Sustainability (LUCID), Lund University, Lund, Sweden; [b]Lund University Centre for Sustainability Studies (LUCSUS), Lund University, Lund, Sweden

This paper analyses spatial conflicts in the iSimangaliso Wetland Park (IWP) in South Africa, a state-led 'development for conservation' project and UNESCO World Heritage Site. With inspiration from Henri Lefebvre's theory on the production of space, it examines dialectical processes of the production of conservation space empirically. Two arenas of conflict: fencing and punitive actions against conservation transgressors are discussed in terms of state power in its relational engagement with local space. Spatial conflicts emerge through tensions between the imposed objectives for the conservation of ecological World Heritage – and the subjective space of users and inhabitants. Market-based modernisation and economic growth strategies, which view land as a commodity, rather than as a social-ecological resource for livelihood generation, perpetuate historical insecurities through the alienation of local people from both land and management practices. Other alienating effects include the socially differentiated effects of new rules of governance, the reshaping of old ethnic identities as a result of envisaged benefits from ecotourism and the imposition of new social-ecological values.

With persistent poverty, accelerating resource extraction and climate change, challenges to conserving biodiversity seem increasingly insurmountable (McShane et al. 2011). Habitat transformation, extinction of species and the decline of animal and plant populations demand urgent action (Adams et al. 2004). Biodiversity conservationists frequently hold that protected areas are the best, if not the only means to adequately protect all elements of biodiversity (i.e. genes, populations and landscapes) (Miller, Minteer, and Malan 2011). Protected areas are traditionally understood as those areas with a minimal human presence and history of alteration (Miller, Minteer, and Malan 2011). However, in those instances when the creation of protected areas causes the foreclosure of future land use options, there are potentially significant economic opportunity costs, and substantial negative effects on local people (Adams et al. 2004). For example, fencing around protected areas hinders access to livelihood resources, such as land for grazing and agriculture. There is increasing concern that global efforts to maintain biodiversity are in conflict with those to reduce poverty (Adams et al. 2004).

In the past several decades a variety of 'win-win' approaches have been introduced all over the world to conserve biodiversity, while also furthering local social and economic development (McShane et al. 2011). In recent years, these efforts have been increasingly connected to a market-based policy agenda for socio-economic development (Brockington and Duffy 2010; Igoe and Brockington 2007; Fairhead, Leach, and Scoones 2012). Critiques of market-based conservation have often focused on its impact on social and social-ecological relations (Bakker 2010; Dressler and Büscher 2008; Sullivan 2006). For example, it is argued that, with the enclosure of commons as commodified land, new property relations increasingly sever customary ties and institutions by placing both under the auspices of self-regulating markets (Dressler and Büscher 2008). Both distributive and procedural social justice is a significant concern in these critiques (Adams and Hutton 2007; Dressler and Büscher 2008; Igoe and Brockington 2007; Redford and Sanderson 2000; West, Igoe, and Brockington 2006). Distributive justice relates to the distribution of costs and benefits associated with protected areas, while procedural justice is connected to the ability to be heard in political processes.

This paper enters these debates with an analysis of conservation conflicts in the iSimangaliso Wetland Park (IWP) in South Africa, a state-led 'development for conservation' project (IWPA 2008). Struggles over conservation space are inter-rogated in terms of what they reveal regarding the politics of protected areas. Taking inspiration from Henri Lefebvre's ideas in *The Production of Space* (1992), the paper examines dialectical processes of the production of conservation space empirically. How have the norms of global biodiversity conservation and national and regional economic development been applied in the IWP's formation? At the local level, the politics of protected area consolidation, expressed in conservation conflicts are analysed. The empirical focus is on cases of conflict over fencing, as well as punitive actions taken against conservation transgressors in the IWP. The impact of dialectical processes of the production of space on social and social-ecological relations is assessed through an embedded case study of the KwaDapha community at Bhanga Nek, Kosi Bay, in the Coastal Forest Reserve Section of the Park. Some implications for conservation – arising out of the imposition of new social-ecological values and the socially differentiated effects of new rules of governance – are suggested. Lefebvre's theories have been developed and applied mostly in the field of urban studies; this study tests their relevance in a rural context.

Study area and methods

The iSimangaliso Wetland Park, South Africa

The iSimangaliso Wetland Park (IWP) covers more than 330,000 hectares, stretching 220 kilometres along the Indian Ocean from Kosi Bay, just below the Mozambican border in the north, to Maphelane south of the St Lucia estuary (DEAT 2009). It encompasses one-third of the KwaZulu-Natal coastline and 9% of the entire coastline of South Africa (DEAT 2009). Its eastern boundary is the Indian Ocean and its western boundary is irregular, incorporating the entire Kosi, Sibaya and St Lucia lake systems, as well as the uMkhuze Game Reserve (Figure 1).

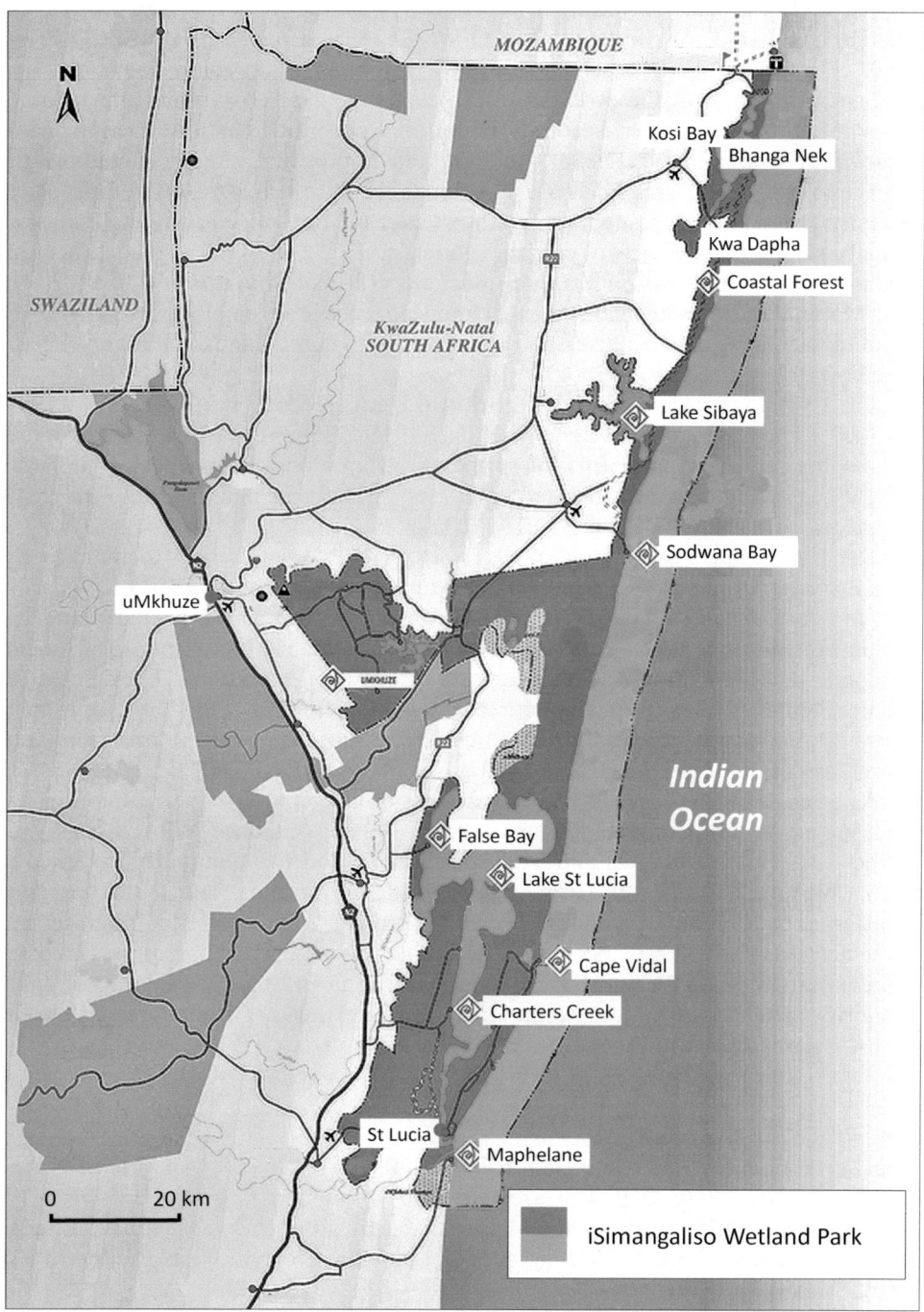

Figure 1. Geographical location of the iSimangaliso Wetland Park.
Source: Adapted from the IWPA (IWPA 2009).

The IWP was listed as South Africa's first UNESCO World Heritage site in 1999. The Park met three of the 10 UNESCO World Heritage criteria (UNESCO WHC 2000). Firstly, the IWP is a representative example of on-going ecological and biological processes in the evolution and development of ecosystems and communities of plants and animals. Secondly, it contains 'superlative natural phenomena or areas of exceptional natural beauty and aesthetic importance'. Lastly, it contains the most important and significant natural habitats for *in situ* conservation of biological diversity, including those containing threatened species of 'outstanding universal value from the point of view of conservation or science'. The IWP also contains four wetlands of international importance under the Ramsar Convention (DEAT 2009). The Maputaland coastal plain is an acknowledged centre of biodiversity, and the Maputaland Centre of Endemism is part of the Maputaland-Pondoland-Albany biodiversity hotspot (IWPA 2008).

The IWP was proclaimed under the World Heritage Convention Act (RSA 1999) in 2000. The IWP effectively consolidated 16 different parcels of previously fragmented land – a patchwork of former proclamations (the earliest going back to 1895); state-owned land; commercial forests; and former military sites – to create an integrated park (IWPA 2008; DEAT 2009).

The iSimangaliso Wetland Park Authority (IWPA) was set-up to manage the Park on behalf of the state (RSA 2000). The major objective of the IWPA is to ensure that the development of the IWP is based on ecotourism as the primary land use option, integrating both the conservation of World Heritage and local socio-economic development. The IWPA reports directly to the national Department of Environmental Affairs, from which it receives its core funding (DEAT 2009). It has a board of nine members, who represent business, traditional councils, land claimants, as well as national, provincial and local government (DEAT 2009).

As a result of historical forced relocations for conservation, the entire park has been subject to competing land claims, with a total of 14 claims (IWPA 2010). Three of these were settled in 1998 and 2002, six in 2007, and five remain to be settled in 2013 (IWPA 2010). In the case of successful land claims, land title has been transferred to claimant communities, with limited user rights under co-management agreements (IWPA 2008; Nustad 2011). The co-management process includes representatives of IWPA and the land claims committee, usually made up of tribal authority members in a given community. Where claims are still to be settled, the IWPA remain the overall managers on behalf of the state (IWPA 2008).

The community of KwaDapha

Local-level research was conducted in KwaDapha, a so-called *tribal authority* area, at Bhanga Nek, Kosi Bay. The area, located within the Coastal Forest Reserve Section of the IWP, comprises four lakes linked by a network of channels. Bhanga Nek lies between the third and biggest lake, kuNhlange, on the west, and the Indian Ocean on the east (Figure 2).

The Kosi Bay Nature Reserve was formally proclaimed in 1987 by the provincial conservation authority, the KwaZulu Bureau of Natural Resources (KBNR) (Kyle 1995). The migration of local people attracted by the infrastructure in KwaNgwanse since the 1970s, accelerated when rumours of the establishment of nature conservation parks in the area were heard in the early 1980s (Mthethwa 2002). Those who

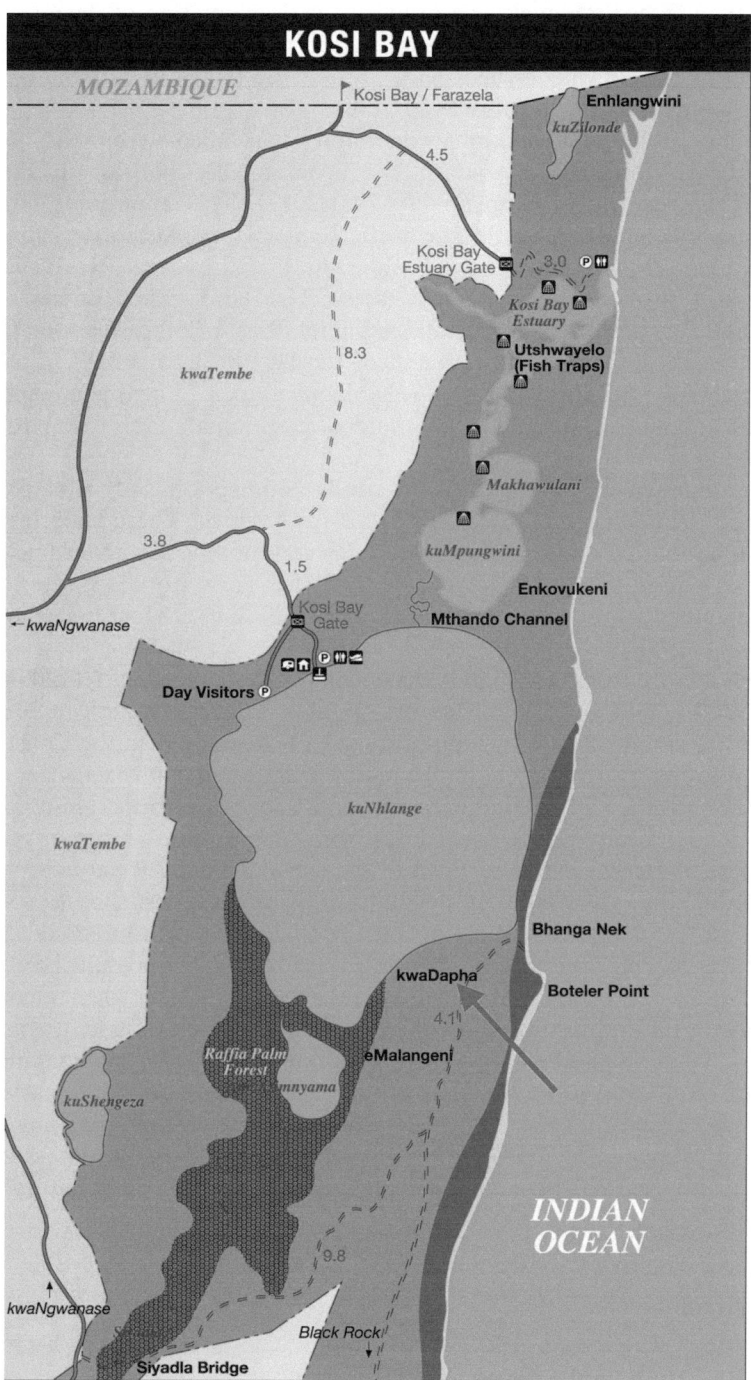

Figure 2. Geographical location of the KwaDapha community at Bhanga Nek, Kosi Bay.
Source: Adapted from the IWPA (IWPA 2009).

stayed resisted forced removals as a result of the proclamation of the reserve (Guyot 2005), but lived under several restrictions from the KBNR (Mthethwa 2002). For instance, local people who owned fields around the banks of kuNhlange were not allowed to plough anymore (Mthethwa 2002).

Since mid-2011, the KwaDapha community has fallen within the uMhlabuya-lingana Local Municipality, one of the economically poorest in the country (uMhlabuyalingana Local Municipality 2011/2012). The community is under the leadership of iNkosi Mabhuda Tembe of the Tembe Tribal Authority, represented by a local iNduna. The area is registered under the Coastal Forest Reserve land claim, which is still to be settled (IWPA 2010). The land claim process has been characterised by continuing contestations and shifting tribal affiliations (Mthethwa 2010). Although under the management of the IWPA, land is held in trust by the iNgonyama Trust, a Zulu tribal trust (uMhlabuyalingana Local Municipality 2011/2012). Title deeds are absent, as the land is communal. Permission to reside in KwaDapha is given by the iNduna.

The community at KwaDapha is physically and economically isolated. KwaNg-wanase is an hour away, accessible only by off-road vehicle. There is a primary school at KwaDapha, but secondary school attendees commute to KwaNgwanase. There is no electrical power, except for solar or petrol/diesel generators in a few households. Water is obtained from wells, or pumped from kuNhlange. Most households have a subsistence garden.

Households are frequently highly dependent on state pensions (R1200 per month in 2013, ≈US$ 135) and Child Support Grants (R260 per month in 2013, ≈US$ 29). These are supplemented by temporary jobs, for example, in the Coast Care and turtle monitoring programmes, implemented through the provincial conservation authority, Ezemvelo KZN Wildlife. Some people have also found temporary jobs at private and community-run tourist camps from time to time. In recent years, illegal tourism developments have burgeoned in the Coastal Forest Reserve Section of the IWP, in KwaDapha, as well as in neighbouring communities, such as eMalangeni. This has been in an apparent effort by local people to exploit tourism demand in the area. Female-headed households often sell resources, such as Zulu beer and palm wine, and reeds for building and maintenance, to supplement their income. Many people rely on natural resources for their livelihoods. For example, it was observed that *ncema* reeds (*Juncus krausii*) were commonly used for mat making and the building of traditional structures. The production of ilala palm (*Hyphaene coriacea*) wine was also observed, as well as the consumption and sale of fish from the coastal area and kuNhlange. The IWPA raises a concern about increasing pressure on such resources inside the IWP, through the depletion and degradation of natural resources in communal areas (IWPA 2008).

Data collection

The local study is based on field research undertaken in the IWP in 2011 and 2012. Interviews focusing on the governance framework of the IWP were carried out with experts from the IWPA. Fourteen semi-structured interviews and two focus group meetings focusing on conservation conflicts were conducted with local informants at KwaDapha. Furthermore, household surveys were carried out with half of the 49 households in KwaDapha, in order to gain information about the socio-economic context and perceptions of the impacts of conservation management on social space. One of the focus group meetings was attended by men and the other exclusively by women, in order to avoid a potential gender gap in the discussions. All interviews, household surveys and focus group meetings were conducted in confidentiality, and the names of the respondents were withheld by mutual agreement. Relevant legislative and policy documents were also collected and analysed. In addition, newspaper articles relating to conservation conflicts in the IWP were reviewed. To improve the reliability of the results, most of the findings were verified by triangulation between different types of sources.

Spatial conflicts in the iSimangaliso Wetland Park

The history of nature conservation in Southern Africa is complex, with conservation often resulting in conflicts with local people (Fabricius 2004). Two arenas where spatial conflicts are visible in the IWP are around fencing and punitive actions taken against local conservation transgressors. For example, tribal authority leaders of the Mbila, Makhasa, Nibela, and Mnqobokazi communities, adjacent to the IWP, have criticised the construction of a fence as potentially limiting their access to natural resources that are considered important for economic and traditional use (Hansen, Ramasar, and Buchanan 2013). Representatives of three of these communities have refused to allow a fence. The fourth community has permitted the erection of a fence, even though the residents knowingly ignore the IWPA's rules for access to the park. A tribal authority representative explained that they were not complying with the IWPA's requests to restrict cattle grazing in the Park, because the authority was not 'listening to them'. Fences between the IWP and adjacent communities have been cut down at various times and locations, according to tribal authority leaders.

Another arena of spatial conflict in the IWP has been the punitive actions taken against conservation transgressors. There have been both civil and criminal cases against local tourism initiatives at KwaDapha, and elsewhere in the Coastal Forest Reserve Section of the IWP (Plate 1).

The applicants in these cases – the Minister of Water and Environmental Affairs, the IWPA and Ezemvelo KZN Wildlife – feared that the IWP would suffer irreparable damage, that it might lose its status as a World Heritage Site, and that the communities which could benefit through controlled management of the park might suffer hardship, unless unlawful occupiers were stopped and evicted before it was too late (Kuppan 2009). The IWPA likened these tourism development initiatives to ecological theft (Kuppan 2009).

A spatial conflict has also emerged with respect to the Kosi Bay Beach Camp, a community-run tourist facility at KwaDapha. An external partner, who had advertised and taken bookings for accommodation at the Camp, was taken to court by the IWPA in 2009. He pleaded guilty to five contraventions of the Protected Areas Act (Savides n.d.). These included his involvement in website advertisements and bookings for other unauthorised developments in the area.

Other punitive actions have been taken against conservation transgressors else-where in the IWP. For example, a bust on 'illegal poacher' boats in the uMfolozi floodplain area took place in 2012 (IWPA 2012). The action was led by Ezemvelo KZN Wildlife staff, together with members of the South African Border Police and the KwaZulu-Natal Airwing of the South African Police Service. The IWPA reported that:

Plate 1. Local tourism development initiatives within the Coastal Forest Reserve Section of the IWP.

> [w]ithin 20 minutes, a total of 28 vessels – most of which were makeshift and did not meet minimum safety requirements – had been seized and loaded onto vehicles. [A] helicopter also kept the ground team informed of possible aggressive reaction from the poachers, but on this occasion none was forthcoming. The team withdrew without incident.

The IWPA CEO stated in reference to this that:

> [t]he large-scale killing of fish and prawns from gill netting has a direct negative effect on the food supply within the estuarine system. It is not sustainable and is literally taking the food out the mouths of legitimate subsistence fishers' families. It can also impact negatively on tourism and jobs (IWPA 2012).

Nevertheless, community members at KwaDapha continue to engage in gill net fishing activities for subsistence. On occasion this has led to intra-community conflict, with some community members having been accused of informing the field rangers of the use of gill nets.

Theorising protected areas as processes of the production of space

Roth (2008) argues that the establishment of protected areas can be understood as a moment of spatial reorganisation resulting from the continual processes of spatial production. She proposes a relational approach to the spatiality of conservation conflict, arguing that space both results from and influences social, political and economic processes. The language of spatial production refers to Henri Lefebvre's best known work *The Production of Space* (1992) – where he insists that space does not merely exist in relation, but is also *produced* in relation (Roth 2008). Lefebvre's work has been applied mostly in the urban planning field (Harvey 2009; Soja 1980), but is flexible enough to be applied in rural contexts (Leary 2009). For example, Neumann (2001) demonstrates that relocations of wildlife and people through British colonial conservation and development plans in Tanzania were fundamental to its construction as a modern nation state.

Lefebvre argues that there exists a dialectical interaction between a society and that society's space – pointing to the contradictory, conflictual and ultimately, political character of the processes of space production. Lefebvre suggests a method for approaching spatial problems, which analyses the contradictions in the utilisation of space by society, and in particular through social customs (Lefebvre and Enders 1976). He draws from Hegel and Marx's dialectical logic, seeing production not only as the creation of material things, but also as an essential part of the reproduction of social relationships (Konzen 2013). Space, then, appears simultaneously as a *material product* resulting from the process of social production (space as product-produced), a *productive force* affecting social production (space as product-producer), and the *physical site* where living bodies interact as a necessary condition for social relations (space as product-medium) (Konzen 2013). Lefebvre's dialectical analysis relies on three elements of a spatial triad – *representations of space*, *spatial practice* and *representational space* (Lefebvre 1992) (Figure 3).

The first category in Lefebvre's (1992) triad is 'Representations of space', or conceived space. This is the ideological space of scientists, planners, urbanists,

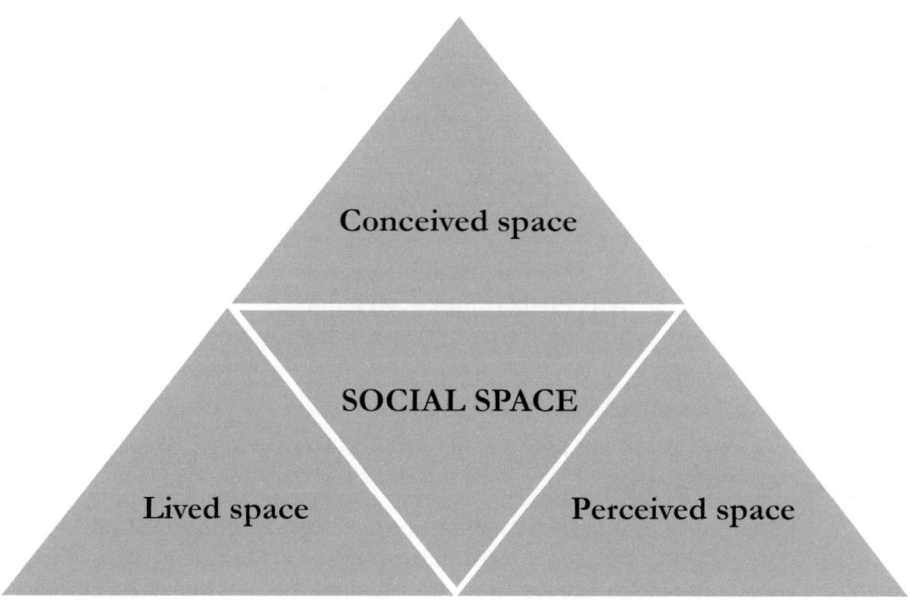

Figure 3. Diagrammatic illustration of Lefebvre's conceptual triad. Social space is produced by dialectical interrelationships amongst conceived space (or representations of space), perceived space (or spatial practice) and lived (or representational) space.

technocratic sub-dividers and social engineers (Lefebvre 1992). This article describes the IWP as the conceived space of politicians and conservation planners. Representations of space, as envisioned through policy documents and conservation plans, are the main subject of discussion. An important aspect of Lefebvre's critique of urban planning relates to the idea of conceived space, for Lefebvre (2003) indicts urban ideology as reductive in its practice (of 'habiting' urban reality). He uses a medical analogy, referring to an urbanist who perceives 'spatial diseases', where space is conceived abstractly as an available void, which must be taken care of so that it can be returned to health.

The second category, 'spatial practice', is empirically observable. It comprises physical interventions that change the materiality of the environment, such as fencing; and the appropriation of material sites by living bodies (Konzen 2013). Here, the focus is on mechanisms for conservation management. Lefebvre distinguishes between *dominated* and *appropriated* space. The former refers to a space transformed – and mediated – by technology, practice, or 'the realization of a master's project' (Lefebvre 1992, 165). Contrastingly, a space appropriated by a group is 'a natural space modified in order to serve the needs and possibilities of [that] group' (Lefebvre 1992).

The last category, 'representational space', or lived space is (subjective) space as directly lived through its associated images and symbols, and hence the space of 'users' and 'inhabitants' (Lefebvre 1992). This is an analysis of place, or what Lefebvre calls 'everyday life', consisting of particular rhythms of being that confirm and naturalise the existence of certain spaces (Thrift 2009).

In what follows, I attempt to assess spatial conflicts in terms of the dialectical interrelationships between *representations of space*, *lived space* and *spatial practice* in the IWP. The analysis starts by describing the political economy of 'development for conservation' in the IWP. *Representations of space* are analysed through broad policy frameworks for regional development in Southern Africa, as well as through legislative documents specific to the implementation of World Heritage conservation in the IWP. These policy and legislative documents are given material effect through *spatial practice* for conservation management. Spatial conflicts emerge through the appropriation of space by local people, in instances where *representations of space* contradict local *representational* (lived) *space*. The analysis assesses spatial conflict in terms of *dominated* and *appropriated* space, focusing on state power in its relational engagement with local space. The dialectical interaction between *spatial practice* and *representational* (lived) *space* is furthermore analysed in terms of changing social and social-ecological relations, arising through the introduction of new values and norms in conserved space.

Representations of space in the iSimangaliso Wetland Park: the political economy of development for conservation

Nature conservation in Southern Africa has always been an important political tool, at least since the advent of European colonisation (Spierenburg and Wels 2006). For example, Carruthers (1995) argues that the proclamation of the Kruger National Park in 1926 was closely linked to the resurgence of Afrikaner nationalism. Representations of space in the IWP can be understood in terms of the political ambitions for the conservation of ecological World Heritage, as well as for national modernisation and economic growth strategies. As a UNESCO World Heritage site, the governance framework of the IWP is partly founded upon normative discourses of global conservation and sustainability. Through the World Heritage Convention Act, a global commitment to the conservation of areas of 'outstanding universal value' has received national legislative support and, in this case, been given effect through the establishment of the IWP.

The IWP illustrates how international sustainability goals and initiatives are integrated, interpreted and operationalised in national modernisation and economic growth policies. Given South Africa's history of dispossession and inequality in terms of access to natural resources, particularly in the context of conservation, the IWPA specifically strives to balance conservation and sustainable development. The policy basis for conservation management in the IWP is the integrated management plan (IMP) (IWPA 2008) – a five-year management plan developed under the World Heritage Convention Act (RSA 1999), along with the National Environmental Management: Protected Areas Act (RSA 2003). The IMP strives to integrate conservation, tourism development, and the local economic development of communities in and adjacent to the Park. However, the Plan is clear that the balance between conservation and locally beneficial economic development is not an equal one. Conservation objectives are prioritised, in order to ensure that World Heritage values are not compromised (IWPA 2008).

Nevertheless, the IMP states that 'economic empowerment and job creation, through appropriate tourism development, is necessary to achieve conservation goals' (IWPA 2008). The IMP strives for the local economic development of historically disadvantaged communities through equity partnerships between the private sector and mandatory community partners (IWPA 2008). An example is the Thonga Beach Lodge and Mabibi community campsite, initiated in 2002 by the IWPA. These have been cited as benchmarks for the development of nature-based tourism partnerships between the private sector and communities (Sunde and Isaacs 2008). This follows what Dressler and Büscher (2008) term a 'hybrid' approach to community-based natural resource management (CBNRM), to denote projects based on private sector investment in 'community based' activities, such as tourism. They argue that this is a sort of 'hybrid neo-liberalism', which merges capitalism and conservation to by-pass the 'subsistence core' of rural livelihoods. A new set of social and social-ecological values and norms are introduced, based upon market-based conceptions of conservation and development, which substitutes direct use of natural resources with indirect alternative forms of economic development (Whande 2010).

Tourism also underlies the rise of the Transfrontier Conservation Area (TFCA) discourse in Southern Africa (Whande 2010). The IWP is a key node of the Lubombo Transfrontier Conservation and Resource Area (LTFCA), a collaborative development project of the governments of Mozambique, South Africa and Swaziland. Although not new, TFCAs have become an important part of a wider context of forms of transnational management of the environment (Duffy 2006). TFCA initiatives create potential benefits in the form of contributing to the maintenance of key ecological functions, sharing management expertise and capacity, better enforcement against poaching, fewer border and customs complexities, and benefits created through nature-based tourism and other types of entrepreneurial ventures (Fakir 2000).

On the other hand, TFCAs have been criticised for allowing a greater degree of centralisation of power and authority over resources and people in the hands of a narrow network of international NGOs; international financial institutions; global consultants on tourism/community conservation; and bilateral donors (Duffy 2006). Whande (2010) shows that top-down virtual mapping, rather than consultation with the local communities, characterises the governance frame of the Greater Limpopo Transfrontier Conservation Area (GLTFCA). He demonstrates a convergence of national tourism interests with protected areas' managerial preference for retaining strong centralist and bureaucratic approaches in the GLTFCA. Duffy (2006) argues that TFCAs, far from being democratic, accountable and transparent forms of environmental management, can often be more accurately viewed as undemocratic, centralising and top-down entities. Thus, while defined as multiple use zones, TFCAs act as windows of exclusion, with implementation heavily reliant on state actors and processes (Whande 2010). Whande (2010) argues that this state-centred model of governance, based on the status of biodiversity as a public good, is a continuation of protectionist and exclusionary approaches to conservation dominant in the first half of the twentieth century. Although the IWP does strive to be a new model for protected area development and management in South Africa – aiming to deliver 'Benefits beyond Boundaries'[1] (IWPA 2008, 11) – the state does indeed have a central role in conservation. This is expressed through the consolidation of the IWP and a strengthened legal framework for the conservation of ecological World Heritage, with the IWPA as the management authority for the state, responsible for implementation of the IMP.

The IWP also forms part of the Lubombo Spatial Development Initiative (LSDI).[2] A principle aim of the LSDI is to generate economic growth by making maximum use of the inherent, but underutilised, tourism potential of the area (IWPA 2008). At a speech at the launch of the LSDI, given in Durban on 6 May 1998, President Nelson Mandela stated:

> [t]he potential of the Lubombo development initiative for tourism and agriculture is truly amazing. Even more remarkable is the extent to which an area of such abundant natural wealth has suffered from neglect. Now that we are all free, our three nations can work together for the development of this region as a whole and realize its true potential.

Within this framework, the IWP is conceptualised as a 'commercial asset that has the potential to help drive the economic revival of a region that was systematically underdeveloped in the past' (IWPA 2008, 3). Economic growth is represented as a cure for regional underdevelopment, the latter being analogous to Lefebvre's conception of 'spatial disease' (2003, 157). As in the case of metropolitan regionalism (Buser 2012), the power of regional economic development has rendered the expression of counter-narratives, such as subsistence resource-based livelihoods, unreasonable. Lived experience is represented as flawed and to be denied in favour of abstract conceptions of economic growth (Buser 2012). This discourse around the necessity of market-based tourism for conservation intersects with the wider global development of a sense of the triumph of neoliberalism, which Swyngedouw (2011) has referred to as 'post-politicisation'. The representation of 'spatial disease' (regional underdevelopment) has enabled state control of large portions of land.

An example is co-management agreements, which have been the predominant approach for reconciling land claims and biodiversity in South Africa (Kepe 2008). Co-management agreements have been signed with nine of the fourteen registered land claims in the IWP. Through co-management agreements, successful land claimants have a share in revenues generated from the conservation area, but do not move back onto the land. This allows protected areas to be kept as conservation land, substituting subsistence resource-based livelihoods with indirect economic development through tourism (Whande 2010). This is a move from seeing land as a social right to property to portraying land as a productive asset (Nustad 2011), consistent with a market-oriented policy paradigm. This approach has led to an erosion of land rights through a land claim settlement policy that views land reform only as a question of the transfer of land (Nustad 2011)

Spatial practice in the iSimangaliso Wetland Park: dominated and appropriated space

Spatial practice is a concept that refers both to the physical environment and empirically observable behaviours – that is, particular social groups' presences, actions, and discourses (Konzen 2013). Simply put, this concept encompasses the way space is perceived and also how people behave in everyday life (Konzen 2013). In the IWP, conservation and preservation measures to preserve ecological integrity and endemism are embodied in spatial practices, such as the use of fencing as a material tool for conservation management. Attempts to control spatial patterns through the establishment of strict protected areas can be understood as an instance of *spatial domination*. The IWP's identification as a UNESCO World Heritage Site raises it above the status of the local territory and, sometimes, beyond the decision-making authority of local people. The strength of the global impetus is reflected in the fact that the IWP's World Heritage Site status privileges certain actors over others, and that conservation goals are prioritised over those of local socio-economic development. This could lead to injustices of 'mis-framing', in which some issues are framed as being primarily of local importance, yet are obliged to compete for resources with issues that are considered to be of international or national importance. An applicable notion here is Nancy Fraser's distinction between two levels of (mis)representation: ordinary political misrepresentation and a higher order concern with frame setting, which appropriates political space at the expense of the poor (Lovell 2007).

For local people, fencing often symbolically represents power relations that lead to their continued exclusion from access to resources, decision-making and co-management. This is illustrated in a media statement released on 4 November 2009, by a committee representing the Bhangazi, Dukuduku, Western Shores, Sokhulu, Mbila, Mdletsheni, KwaJobe and Triangle communities (adjacent to the IWP). They expressed their 'wish to bring to the attention of the world and government the concern that their rights to access land for grazing, cropping and hunting are severely curtailed' (Savides 2009). For people living in the IWP, fencing means that they are enclosed with dangerous wildlife, frequently with negative impacts upon their livelihood strategies. At KwaDapha, there was a constant refrain against the destruction of subsistence gardens by hippopotami (*Hippopotamus amphibious*) and vervet monkeys (*Chlorocebus pygerythrus*).

The imposition of new rules of governance for conservation management in the IWP also suggests spatial domination. This is perhaps most obvious when these new rules constrain the economic and social development activities of – and opportunities for – local people. Residents often perceived the lack of jobs in KwaDapha as a result of the restrictions on local tourism development initiatives. These perceptions were evident in discussions about the Kosi Bay Beach Camp. There had been a decrease in the number of tourists staying at the Camp, with community members arguing that this was a result of difficulties with advertising. Even official maps for the IWP are silent about the Camp, although responsibility for day-to-day management had been transferred from Ezemvelo KZN Wildlife to the community in 2001. People also had negative attitudes toward the IWPA because of sanctions elsewhere in their everyday lives. The iNduna stated that '[w]e are not free in this area'. Another respondent maintained that: '[a]fter iSimangaliso came in 1999 they put sanctions on us. Life was better before. Now there are sanctions even in the lake. People can't renovate their houses, can't fish on the lake'. Views like the following were expressed:

> [w]e want the government to intervene to build big lodges for people to have jobs at KwaDapha. People won't then have a problem with permits or sanctions. [. . .]. We have submitted an application to develop a 4-star diving lodge where the community tented camp currently is and to upgrade the Kosi Bay Beach Camp. Then iSimangaliso will find it easier to work with communities. If iSimangaliso doesn't stop development, they will find it easier to work with the community.

On the other hand, local people who have been politically and economically marginalised from development processes, often turn to non-confrontational forms of everyday resistance (Li 2007; Neumann 2002; van Wyk 2003). Here people resist intrusions on their autonomy through what Scott (1985) has called 'weapons of the weak' – where resistance often takes the form of passive non-compliance, subtle sabotage and quiet evasion. These forms of everyday resistance are 'informal, often covert and concerned largely with immediate, de facto, gains' (32). Such practices can be understood in terms of Lefebvre's concept of *spatial appropriation*, as they are concerned with modifying space in order to serve the needs and aspirations of local people. In the IWP, local people continue to enter to gather natural resources according to their own schedules, knowingly ignoring the IWPA's rules for access (Hansen, Ramasar, and Buchanan 2013). Community members at KwaDapha engage in gill net fishing activities for subsistence.

In other cases of everyday forms of non-confrontational resistance, there is superficial compliance with an arranged (but unauthorised) situation, allowing people to go about their daily tasks without open conflict (van Wyk 2003). This is illustrated, for example, through sometimes obstructionist relationships between the IWPA and land claims committees. Members of the Mnqobokazi community were not observing the rules and regulations established in their co-management agreement, which had not yet been formally signed (Hansen, Ramasar, and Buchanan 2013). One reason for non-compliance with official rules may be the difference between cultural norms and the imposed policy and legislative framework of the IWP. In other words, spatial conflicts emerge when *representational* (lived) *space* (the subjective space of users and inhabitants) conflicts with *representations of space* (the imposed conservation space of political ambitions and conservation planners). Interviewees involved in tourist camp developments stated that they believed they had gone through the necessary channels for authorisation, receiving the go-ahead from the local iNduna and the owners of the land, the iNgonyama Trust. One person was quoted in a newspaper report, stating that '[w]e believe that the court was wrong to rule against us. We followed all the relevant channels before we started building' (Sapa 2009).

Social space in the iSimangaliso Wetland Park
'*iSimangaliso has stolen this area*'. *Exclusion, inclusion and new social relations*

In the IWP, management choices, decision-making structures, and policies in support of conserving a World Heritage Site, are guided by discourses around global conservation. Although local needs are acknowledged, the World Heritage status of the IWP means that some decisions have been taken beyond the bounds of the local area (Hansen, Ramasar, and Buchanan 2013). This is attested to by the research finding that residents of KwaDapha did not participate in the designation of the Kosi Bay area as part of a World Heritage Site. It was said in a focus group meeting that 'iSimangaliso has stolen this area. They [the IWPA] were supposed to ask our permission to declare this a World Heritage Site. We are confused because we haven't even seen the papers that say this is a World Heritage Site'. Local people also expressed the perception that they have no voice in future plans for the Kosi Bay area. A community based development committee had submitted an application to the IWPA to develop a diving lodge, in partnership with an external investor. However, they had not received a reply since submitting their application in 2009. They believed that this was because the IWPA had other plans for the area. One community member stated that '[w]e do have our own plans, but our plans do not matter so much because they [the IWPA] have their own plans'. Nancy Fraser (2010) suggests that injustices at intersecting scales can lead to the social exclusion of the global poor. A question of justice arises not from simply looking at the local level, but rather through understanding the interplays of power taking place at the intersection of several levels (Hansen, Ramasar, and Buchanan 2013).

The restructuring of rules and authority over the access, use and management of resources can have alienating effects (Fairhead, Leach, and Scoones 2012). van Wyk (2003) found that in Maputaland, only certain men had access to the privileged domain of negotiation with the state by virtue of their claim to be traditionally sanctioned representatives of local people. Often they used these negotiations to further their own private political and economic interests. Sunde and Isaacs (2008) report that many people from the Mabibi community are unaware of the potential benefits flowing to them from the Thonga Beach Lodge and the Mabibi community campsite. They cite the hierarchical structure of the local tribal authority as one of the reasons for this (Sunde and Isaacs 2008). Mthethwa argued in 2002 that the reshaping of old ethnic identities and the local leadership's mobilisation of history in Maputaland is inspired by the envisaged economic benefits to be derived from the advent of eco-tourism (2002).

Research has also found that other government departments have been effectively fenced out through the dominance of the IWPA in governing the area. Conducting research in the Nibela, Makhasa and Mnqobokazi communities, adjacent to the

IWP, Buchanan (2011) found a lack of interaction between the IWPA and municipalities. Municipal officials from the Big Five False Bay Local Municipality stated that they had never had contact with the IWPA. The IWPA, for its part, stated that the roles of the authority and the municipalities were separate and unrelated (Buchanan 2011). This has important implications for service delivery at the local level. Under the South African Constitution, local government has considerable autonomy and the responsibility to promote social and economic development (Frödin 2011). Municipalities also provide the linkages to the provincial and national departments that are responsible for other services, such as health care and education (RSA 1998). In reference to a perceived lack of service delivery at KwaDapha, one community member said: '[i]t's a new South Africa, but we are still living like in Apartheid times. Nothing is happening, we are actually neglected. We are living in a World Heritage Site, but we are still neglected'. Oviedo and Puschkarsky (2012) note that human rights violations in World Heritage sites have been perpetuated through policies of avoiding or minimising basic service provision, such as healthcare for communities forced to remain isolated and without access to infrastructure.

'We are sleeping with the hippos'. New social-ecological relations

A strategy that aims to conserve nature through the establishment of strictly bounded protected areas posits a fundamental distinction between humanity and nature. This view is problematic for many reasons. Firstly, protected areas are surrounded by conflicts, including conflicts with local people over access to resources in protected areas. Issues of distributive social justice are raised when asking the questions: 'conservation at whose expense?', and 'for whom?' In the case of the IWP, social justice is a key concern in the implementation gap between the stated policy aims of the local economic development of historically disadvantaged communities, and the alienation of local people through conservation management.

Secondly, the idea that nature and humans are fundamentally opposed is a Western idea arising out of environmental change and a romantic longing for a 'nature' untransformed by industrialisation (Lefebvre and Enders 1976). Imposing the conservation agenda of enclosed protected areas that are separate from human activity raises questions around the dominance of Western ideologies and value systems, addressed by many post-colonial theorists (Fairhead, Leach, and Scoones 2012; Li 2007; Neumann 2002). This is not to suggest a naive view of local people as stewards of the natural environment. Indeed, field research at KwaDapha shows a strong desire for modernist development. Local people often engage in harmful environmental practices, for example the use of gill nets for fishing. The point is that the imposition of firmly bounded protected areas has profoundly alienating effects, in terms of both distributive and procedural justice – the former related to the material distribution of costs and benefits, and the latter related to democratic participation in conservation management.

A related idea is that of green grabbing (Fairhead, Leach, and Scoones 2012), which builds on well-known histories of colonial and neo-colonial resource alienation in the name of the environment. Green grabbing nevertheless constitutes new ways of appropriating nature, through novel forms of valuation, commodification and markets. In the case of the IWP, nature has been commoditised through a market-based conservation approach. Here the IWP is viewed as a commercial asset, rather than a social-ecological resource for livelihood generation. One example of the effect of this at the local level is on subsistence agriculture. A major concern here was, in a women's focus group meeting, where it was stated that: '[n]ow no one is farming'. At another meeting, residents said that '[l]ife is difficult. We don't know what we can do. They [the IWPA] don't allow us to even renovate our households. Rules must go with development. Rules are so strict for people, but there is no development'.

Further, the idea that we can address environmental change by reinforcing the conceptual dualism between humans and nature has been the subject of much critique (Adams and Hutton 2007; Castree 1995; Fitzsimmons 1989; Redford and Sanderson 2000; West, Igoe, and Brockington 2006). A good deal of the literature cited here relates to the people-parks debate (Adams and Hutton 2007; Redford and Sanderson 2000; West, Igoe, and Brockington 2006). The social impacts of protected areas have been recognised by conservation planners since the 1980s (Adams et al. 2004). However the question of whether it is possible to combine poverty elimination and biodiversity conservation relates to a more general debate about the environmental dimensions of development (Adams et al. 2004).

In the case of the IWP, there is ample evidence that conservation management has reinforced a nature-society dualism, evidenced by, for example, an increased intensity of conflict between local people and wildlife. During the household surveys conducted at KwaDapha, most people reported that they had experienced difficulties with what they called 'nature's problem'. Older respondents in particular explained that they had detected an increase in forest cover over the preceding 10 or so years. They viewed this as a negative impact. In a women's focus group meeting it was said that 'we are sleeping with the hippos'.

Conclusions: the social production of conserved nature

The analysis of the policy framework of the IWP in terms of Lefebvre's (1992) representations of space has shown that the Park is conceived as a tool for market-based modernisation and economic growth. The view here is of land as a commodity, rather than as a social-ecological resource for livelihood generation. Spatial conflicts emerge where space is dominated through state led conservation management, and local communities in turn appropriate space for their livelihoods. In this regard, analysing dialectical spatial practices – both in terms of dominated and appropriated space – provides some insights to questions of social justice, through a focus on state power in its relational engagement with local space. This has important implications, as conflicts over land use in conservation areas juxtapose efforts to restore local land and resource rights against national and global interest in conservation (Whande 2010). In the IWP, conservation strategies perpetuate historical insecurities through the alienation of local communities from land, as well as management practices. New social and social-ecological relations arise, in which local people are excluded from meaningful participation in both formulating ideas about the future management of their land and physically from the land itself. In an urban context, Lefebvre (2003, 148) argued for an alternative in the form of a 'politicisation of urban issues', where democratic processes support rather than deny a vibrant politics of contestation. In the context of the IWP, a re-politicisation of conservation space is called for.

The paper uses Lefebvre's theoretical insights in a rural context, through an analysis of dialectical processes of the production of conservation space. Referring to

Lefebvre's conceptual triad, the paper has shown that *representations of space* in the IWP can be understood as the political ambitions for the conservation of ecological World Heritage, as well as for regional and national modernisation and economic growth strategies. Similar to the case of metropolitan regionalism discussed by Buser (2012), even in the rural context processes may be observed where the power of regional economic development has rendered the expression of counter-narratives, such as subsistence resource-based livelihoods, 'unreasonable'. These representations of space are embodied in *spatial practice*, physical interventions that change the materiality of the environment and the appropriation of sites by living bodies.

Space is *dominated* through attempts to control the spatial patterns of residents both adjacent to and in the IWP, exemplified in this paper through examples of fencing and punitive actions against conservation transgressors. In reaction, local people who have been politically and economically marginalised from development processes resist intrusions on their autonomy through 'weapons of the weak' (Scott 1985). Such practices can be understood in terms of Lefebvre's concept of *spatial appropriation*, as they are concerned with modifying space in order to serve the needs and possibilities of local people. It has also been shown that local people continue to follow their own social conventions and norms, in preference to new policy and legislative frameworks imposed upon them from the IWPA.

Spatial conflicts arise through these tensions between the subjective space of users and inhabitants (*representational* (lived) *space*) and the enforced objectives for the conservation of Ecological World Heritage. Local people involved in court cases for illegal tourism development initiatives at KwaDapha argued that they had gone through all the necessary channels for authorisation, receiving the go-ahead from the local iNduna and the owners of the land, the iNgonyama Trust. In turn, local lived space is impacted upon through new rules of governance, often leading to exclusion in decision-making and other alienating effects, including the reshaping of old ethnic identities and the imposition of new social-ecological relations on local communities.

It has also been shown that Lefebvre's work may be relevant for analysing processes of global environmental governance. This may be particularly important in light of recent calls for a return to planning (for example, with regard to the politics of climate change) (Giddens 2009). As shown here, the implementation of global conservation through a market-based approach is particularly problematic where there is structural inequality, such as in South Africa.

Acknowledgements

The author gratefully acknowledges the support of the LUCID Research School for comments and discussions on earlier versions of this paper. The author alone is of course responsible for any shortcomings.

Notes

1. Protected Areas: Benefits beyond Boundaries' was the theme of the Fifth World Parks Congress, held in Durban, South Africa, in 2003. An important outcome of the Congress was the 'Durban Accord' (IUCN 2003), which called for an innovative approach to protected areas and their role in broader conservation and development agendas, emphasising a synergy between conservation and sustainable development. The IWP has

been 'applauded for its pioneering work in integrating World Heritage conservation into regional development – and delivering "Benefits beyond Boundaries"' (IWPA 2008, 11).
2. This is the overarching development project initiated by the governments of Mozambique, South Africa and Swaziland. The LTFCA is a core component of this project.

Note on contributor

Melissa Hansen is a PhD Candidate in Sustainability Science at the Lund University Centre for Sustainability Studies (LUCSUS), Sweden. She is also a member of the LUCID Research Programme, which aims at creating new and unique synergies across natural and social sciences, in order to develop integrated theories and methods for addressing complex sustainability issues. Her research interests include the tensions arising from market-based conservation, specifically in terms of social justice. She is also interested in the politics of protected areas, struggles over conservation space, and dialectical processes of the production of space.

References

Adams, William M., and Jon Hutton. 2007. "People, Parks and Poverty: Political Ecology and Biodiversity Conservation." *Conservation and Society* 5 (2): 147–183. http://www.conservationandsociety.org/text.asp?2007/5/2/147/49228.

Adams, William M., Ros Aveling, Dan Brockington, Barney Dickson, Jo Elliott, Jon Hutton, Dilys Roe, Bashkar Vira, and William Wolmer. 2004. "Biodiversity Conservation and the Eradication of Poverty." *Science* 306 (5699): 1146–1149. doi:10.1126/science.1097920.

Bakker, Karen. 2010. "The Limits of 'Neoliberal Natures': Debating Green Neoliberalism." *Progress in Human Geography* 34 (6): 715–735. doi:10.1177/0309132510376849.

Brockington, Dan, and Rosaleen Duffy. 2010. "Capitalism and Conservation: The Production and Reproduction of Biodiversity Conservation." *Antipode* 42 (3): 469–484. doi:10.1111/j.1467-8330.2010.00760.x.

Buchanan, Kent. 2011. *Disjoined Action in a Conjoined World: An Analysis of Human Development Governance in Rural KwaZulu Natal, South Africa.* Lund: Lund University Centre for Sustainability Studies, Lund University.

Buser, Michael. 2012. "The Production of Space in Metropolitan Regions: A Lefebvrian Analysis of Governance and Spatial Change." *Planning Theory* 11 (3): 279–298. doi:10.1177/1473095212439693.

Carruthers, E. Jane. 1995. *Game Protection in the Transvaal 1846 to 1926, Argiefjaarboek vir Suid-Afrikaanse geskiedenis.* South Africa: Government Printer.

Castree, Noel. 1995. "The Nature of Produced Nature: Materiality and Knowledge Construction in Marxism." *Antipode* 27 (1): 12–48. doi:10.1111/j.1467-8330.1995.tb00260.x.

DEAT. 2009. *Fifteen Years: A Review of the Department of Environmental Affairs and Tourism 1994 –2009.* Pretoria: Chief Directorate Communications, Department of Environmental Affairs and Tourism.

Dressler, Wolfram, and Bram Büscher. 2008. "Market Triumphalism and the CBNRM 'Crises' at the South African Section of the Great Limpopo Transfrontier Park." *Geoforum* 39 (1): 452–465. doi:10.1016/j.geoforum.2007.09.005.

Duffy, Rosaleen. 2006. "The Potential and Pitfalls of Global Environmental Governance: The Politics of Transfrontier Conservation Areas in Southern Africa." *Political Geography* 25 (1): 89–112. doi:10.1016/j.polgeo.2005.08.001.

Fabricius, Christo. 2004. *Rights, Resources and Rural Development: Community-Based Natural Resource Management in Southern Africa.* London: Earthscan.

Fairhead, James, Melissa Leach, and Ian Scoones. 2012. "Green Grabbing: A New Appropriation of Nature?" *The Journal of Peasant Studies* 39 (2): 237–261. doi:10.1080/03066150.2012.671770.

Fakir, Saliem. 2000. *Transfrontier Conservation Areas: A New Dawn for Eco-Tourism, or a New form of Conservation Expansionism?* Pretoria: IUCN South Africa.

Fitzsimmons, Margaret. 1989. "The Matter of Nature." *Antipode* 21 (2): 106–120. doi:10.1111/j.1467-8330.1989.tb00183.x.

Fraser, Nancy. 2010. "Injustice at Intersecting Scales: On 'Social Exclusion' and the 'Global Poor'." *European journal of social theory* 13 (3): 363–371. doi:10.1177/1368431010371758.

Frödin, Olle. 2011. "Generalised and Particularistic Thinking in Policy Analysis and Practice: The Case of Governance Reform in South Africa." *Development Policy Review* 29 (Suppl 1): S179–S198.

Giddens, Anthony. 2009. *Politics of Climate Change*. Cambridge: Polity.

Guyot, Sylvain. 2005. "Political Dimensions of Environmental Conflicts in Kosi Bay, South Africa: Significance of the New Post-Apartheid Governance System." *Development Southern Africa* 22 (3): 441–458. doi:10.1080/14797580500252985.

Hansen, Melissa, Vasna Ramasar, and Kent Buchanan. 2013. "Localizing Global Environmental Governance Norms: Implications for Justice." In *Governance for Justice and Environmental Sustainability*, edited by Merle Sowman and Rachel Wynberg. Cape Town: Routledge.

Harvey, David. 2009. *Social Justice and the City, Geographies of Justice and Social Transformation Series*. Athens, GA: University of Georgia Press.

Igoe, James, and Dan Brockington. 2007. "Neoliberal Conservation: A Brief Introduction." *Conservation & Society* 5 (4): 432–449. http://www.conservationandsociety.org/text.asp?2007/5/4/432/49249.

IWPA (isimangaliso Wetland Park Authority). 2008. *Integrated Management Plan*. The Dredger Harbour, St Lucia: IWPA.

IWPA (isimangaliso Wetland Park Authority). 2009. "iSimangaliso Wetland Park: Maps 2009". Accessed February 4, 2013. http://www.isimangaliso.com/index.php?maps.

IWPA (isimangaliso Wetland Park Authority). 2010. "Presentation by the iSimangaliso Wetland Park Authority." Paper read at The fourth People and Parks National Conference, at Cape Vidal, iSimangaliso Wetland Park, South Africa, August 29–September 1.

IWPA (isimangaliso Wetland Park Authority). 2012. *The Big Boat Bust and Rhino Outrage!* edited by IWPA. The Dredger Harbour, St Lucia: IWPA.

IUCN. 2003. "Message of the Vth IUCN World Parks Congress to the Convention on Biological Diversity." *Journal of International Wildlife Law & Policy* 6 (3): 277. doi:10.1080/13880290390437300.

Kepe, Thembela. 2008. "Land Claims and Comanagement of Protected Areas in South Africa: Exploring the Challenges." *Environmental Management* 41 (3): 311–321. doi:10.1007/s00267-007-9034-x.

Konzen, Lucas P. 2013. *Norms and Space: Understanding Public Space Regulation in the Tourist City*. Milan: Department of Law, University of Milan.

Kuppan, Irene. 2009. "Tear it Down – or Pay." *The Daily News*, November 24, 2009.

Kyle, Robert. 1995. *Kosi Bay Nature Reserve South Africa: Information Sheet for the Site Designated to the List of Wetlands of International Importance Especially as Waterfowl Habitat*, edited by South African Wetlands Conservation Programme. Pretoria: Department of Environmental Affairs and Tourism (DEAT).

Leary, Michael E. 2009. "The Production of Space through a Shrine and Vendetta in Manchester: Lefebvre's Spatial Triad and the Regeneration of a Place Renamed Castlefield." *Planning Theory & Practice* 10 (2): 189–212. doi:10.1080/14649350902884573.

Lefebvre, Henri. 1992. *The Production of Space*. Oxford: Wiley-Blackwell.

Lefebvre, Henri. 2003. *The Urban Revolution*. Minneapolis and London: University of Minnesota Press.

Lefebvre, Henri, and Michael J. Enders. 1976. "Reflections on the Politics of Space." *Antipode* 8 (2): 30–37. doi:10.1111/j.1467-8330.1976.tb00636.x.

Li, Tanya M. 2007. *The will to Improve: Governmentality, Development, and the Practice of Politics*. Durham: Duke University Press.

Lovell, Terry. 2007. "Introduction." In *(Mis)recognition, Social Inequality and Social Justice: Nany Fraser and Pierre Bourdieu*, edited by Terry Lovell, 1–16. London: Routledge.

McShane, Thomas O., Paul D. Hirsch, Tran C. Trung, Alexander N. Songorwa, Ann Kinzig, Bruno Monteferri, David Mutekanga, Hoang V. Thang, et al. 2011. "Hard Choices:

Making Trade-Offs Between Biodiversity Conservation and Human Well-Being." *Biological Conservation* 144 (3): 966–972. doi:10.1016/j.biocon.2010.04.038.

Miller, Thaddeus R., Ben A. Minteer, and Leon-C. Malan. 2011. "The New Conservation Debate: The View from Practical Ethics." *Biological Conservation* 144 (3): 948–957. doi:10.1016/j.biocon.2010.04.001.

Mthethwa, Dingani. 2002. "The Mobilization of History and the Tembe Chieftaincy in Maputaland: 1896–1997." Masters diss., University of Natal, Durban.

Mthethwa, Dingani. 2010. "Two Bulls in One Kraal: Local Politics, 'Zulu History,' and Heritage Tourism in Kosi Bay." In *Zulu Identities: Being Zulu, Past and Present*, edited by Benedict Carton, John Laband, and Jabulani Sithole, 499–514. Columbia University Press.

Neumann, Roderick P. 2001. "Africa's 'Last Wilderness': Reordering Space for Political and Economic Control in Colonial Tanzania." *Africa: Journal of the International African Institute* 71 (4): 641–665. doi:10.3366/afr.2001.71.4.641.

Neumann, Roderick P. 2002. *Imposing Wilderness: Struggles Over Livelihood and Nature Preservation in Africa, California Studies in Critical Human Geography*. Berkely, CA: University of California Press.

Nustad, Knut G. 2011. "Property, Rights and Community in a South African Land-Claim Case." *Anthropology Today* 27 (1): 20–24. doi:10.1111/j.1467-8322.2011.00784.x.

Oviedo, Gonzalo, and Tatjana Puschkarsky. 2012. "World Heritage and Rights-Based Approaches to Nature Conservation." *International Journal of Heritage Studies* 18 (3): 285–296. doi:10.1080/13527258.2012.652146.

Redford, Kent H, and Steven E. Sanderson. 2000. "Extracting Humans from Nature." *Conservation Biology* 14 (5): 1362–1364. doi:10.1046/j.1523-1739.2000.00135.x.

Republic of South Africa (RSA). 1998. *Local Government: Municipal Structures Act (Act 117 of 1998)*. Cape Town: Government Gazette.

Republic of South Africa (RSA). 1999. *World Heritage Convention Act (Act no. 49 of 1999)*. Notice 20717 of 414. Cape Town: Government Gazette.

Republic of South Africa (RSA). 2000. *Establishment of the Greater St Lucia Wetland Park and Authority*. In *Notice 4477 of 2000*, edited by Department of Environmental Affairs and Tourism. Pretoria: Government Gazette.

Republic of South Africa (RSA). 2003. "National Environmental Management: Protected Areas Act (Act 57 of 2003)." In *Notice No. 99. Government Gazette No. 35021 of 8 February 2012*, edited by Republic of South Africa. Cape Town: Government Gazette.

Roth, Robin J. 2008. "'Fixing' the Forest: The Spatiality of Conservation Conflict in Thailand." *Annals of the Association of American Geographers* 98 (2): 373–391. doi:10.1080/00045600801925557.

Sapa. 2009. "Builders Brush off Order." Accessed December 29. *iafrica.com*.

Savides, Dave. 2009. "Threat to 'shut down tourism'." *Zululand Observer*, November 10, 2009.

Savides, Dave. n.d. "Landmark Verdict for Wetland Park." *Zululand Observer*.

Scott, James C. 1985. *Weapons of the Weak: Everyday forms of Peasant Resistance*. New Haven: Yale University Press.

Soja, Edward W. 1980. "The Socio-Spatial Dialectic." *Annals of the Association of American Geographers* 70 (2): 207–225. doi:10.1111/j.1467-8306.1980.tb01308.x.

Spierenburg, Marja, and Harry Wels. 2006. "'Securing Space': Mapping and Fencing in Transfrontier Conservation in Southern Africa." *Space and Culture* 9 (3): 294–312. doi:10.1177/1206331206289018.

Sullivan, Sian. 2006. "The Elephant in the Room? Problematising 'New' (Neoliberal) Biodiversity Conservation." *Forum for Development Studies* 33 (1): 105–135. doi:10.1080/08039410.2006.9666337.

Sunde, Jackie, and Moenieba Isaacs. 2008. "Marine Conservation and Coastal Communities: Who Carries the Costs? A Study of Marine Protected Areas and Their Impact on Traditional Small-scale Fishing Communities in South Africa." In *SAMUDRA Monograph*, edited by K. G. Kumar, 1–9. Chennai: International Collective in Support of Fishworkers.

Swyngedouw, Erik. 2011. "Interrogating Post-Democratization: Reclaiming Egalitarian Political Spaces." *Political Geography* 30 (7): 370–380. doi:10.1016/j.polgeo.2011.08.001.

Thrift, Nigel. 2009. "Space: The Fundamental Stuff of Geography." In *Key Concepts in Geography*, 2nd ed., edited by Nicholas Clifford, Shaun Holloway, and Gill Valentine, 95–107. London: Sage.

uMhlabuyalingana Local Municipality. 2011/2012. *IDP Review*, edited by Office of the Municipal Manager. KwaNgwanase: uMhlabuyalingana Local Municipality.

UNESCO WHC. 2000. *Report of the 23rd Session of the Committee*. Marrakesh: UNESCO World Heritage Committee.

van Wyk, Ilana. 2003. "Elephants are Eating Our Money: A Critical Ethnography of Development Practice in Maputaland, South Africa." MA diss., Anthropology and Archeology, University of Pretoria, Pretoria.

West, Paige, James Igoe, and Dan Brockington. 2006. "Parks and Peoples: The Social Impact of Protected Areas." *Annual Review of Anthropology* 35 (1): 251–277. doi:10.1146/annurev.anthro.35.081705.123308.

Whande, Webster. 2010. "Windows of Opportunity or Exclusion? Local Communities in the Great Limpopo Transfrontier Conservation Area, South Africa." In *Community Rights, Conservation and Contested Land: The Politics of Natural Resource Governance in Africa*, edited by Fred Nelson, 147–173. London: Earthscan.

Balancing (re)distribution: Franco-Mauritians landownership in the maintenance of an elite position

Tijo Salverda

Human Economy Programme, Faculty of Humanities, University of Pretoria, Pretoria, South Africa

Sugarcane, once a vital component of the Mauritian economy, now makes up less than 10% of GDP. The land on which it grows (and grew), however, remains essential to the island's politico-economic power balance. Large tracts of land are still owned by Franco-Mauritians, the island's white former colonial elite, thus leading to an ambiguous relationship between landowners and the government in postcolonial Mauritius. This article argues that both resentment and collaboration contribute to the consolidation of the Franco-Mauritian elite position. Pressure from the state may have compelled Franco-Mauritians to redistribute some of their land, but this 'pay-off' hardly jeopardised their elite position. Their striking a balance between opposing redistribution and giving in to government demands is key to explaining how landed (white) elites are, in the absence of (state) violence, able to maintain their position long after the end of colonialism.

Mauritius, nowadays a renowned tourist destination, relied almost solely on sugar production during colonial times. Most of the land on which the sugarcane was grown was owned by Franco-Mauritians, the white (former) colonial elite of the island. They were given large tracts of the island's uninhabited land at the start of the French colonial project, which they used for the cultivation of sugarcane. This helped them to accumulate wealth and to remain at the peak of the socio-economic pyramid relatively easily until the 1930s. From then, the apex position of the Franco-Mauritians began experiencing challenges with the emancipation of the other sections of the Mauritian population and democratisation of the political process, culminating with the independence of Mauritius in 1968 as its apogee.[1] In this process, Mauritians of Hindu origin, who make up about half of the population, became the dominant force within the political domain and state apparatus. Nevertheless, more than 40 years after independence Franco-Mauritians, who are currently estimated to number around 10,000, making up slightly less than 1% of the total population,[2] still control large parts of the island's agricultural land, and they remain an elite. They are the island's wealthiest community and no Franco-Mauritian tends to be found among the working class (Eriksen 1998, 62). Today, Franco-Mauritians have diversified their economic interests, while mergers and buy-outs in the sugar industry have also led to a concentration of Franco-Mauritians possessing agricultural land. The origin of the community's wealth, nonetheless, is undeniably tied to the land and associated with colonial injustices. This explains why land remains important to understanding the position of the Franco-Mauritians.

White landownership in former European colonies, especially in Africa, is contentious. In Zimbabwe, white farmers have been driven off the land, while in South Africa as well, the redistribution of land remains a prominent item on the political agenda. The Mauritian case, and thus Franco-Mauritian landownership, differs from mainland Africa, because Mauritius was uninhabited before the arrival of the European colonists, which gives white landownership a different meaning than on the continent.[3] However, the privileges enjoyed by the Franco-Mauritians during the colonial period closely resembled the ones that white populations elsewhere in Africa benefited from. Landownership was one of the most, if not the most, prominent of these privileges. This advantage of ownership shapes memories of colonial injustices and the heritage of a system favouring Franco-Mauritians up till today, and makes the land question in Mauritius and the sentiment around this question comparable to white landownership in Africa more generally. It is important to note that white African farmers do not always enjoy the same elite position as Franco-Mauritians, even if they tend to be better off than the majority of Africans. The Franco-Mauritian case, then, is not only about being white but also about being an elite. In this sense, the Franco-Mauritian case is comparable not only to African whites but also to elites more generally as it helps to explain the functioning of land (and other resources) in the maintenance of an elite position. The Creole elite of Sierra Leone (Cohen 1981) and the Indian Parsis (Luhrmann 1996), for example, had little control over tangible resources. In the case of the Parsis, who were an administrative elite during the colonial period, the decolonisation of India led to their demise. They were more easily outdone by newly emerged elites because they did not have assets that 'protected' them against rival counter-forces. Franco-Mauritians, however, could 'mobilise' their lands.

Agricultural products, sugar in the case of Mauritius, obviously generate wealth for the landowners. This article, however, is concerned with the functioning of land in the relationship between the landowners and opposing groups and/or institutions, the Mauritian postcolonial government in this case. While the generation of wealth is related to this process, it is not the core of my focus. Rather, the article aims to illustrate how the ambiguity of Franco-Mauritian landownership in the relationships between landowners and the government contributes to the maintenance of an elite position. To better understand land in contemporary Mauritius, I will first address the historical power base of the Franco-Mauritian community and the impact of this heritage in the postcolonial period. Second, I will address the present-day role of land with respect to the maintenance of the Franco-Mauritian elite position. For many Mauritians, land symbolises the white elite's unequal economic power and unfair colonial heritage. Mauritian politicians, who are predominantly of Hindu origin, often criticise Franco-Mauritian landownership and also challenge Franco-Mauritians to (re)distribute parts of their land. The manner in which Franco-Mauritians face these challenges is key to the understanding of their elite position, in terms of wealth as well as a shared sense of belonging among members of this community. In the negotiations on the restructuring of the sugar industry, for example, the government pressured Franco-Mauritian sugar estates to redistribute part of their land to retrenched labourers. Franco-Mauritian landowners opposed any restitution, and garnered the support of most members of the Franco-Mauritian community. Yet, they eventually had to concede to the majority of the government's demands. While Franco-Mauritians may have had to distribute some of their land, this 'pay-off' nonetheless gave them some respite and hardly jeopardised their position. This striking a balance between opposing redistribution and giving in to government demands may help to prevent substantial decline for a long time to come – as elaborated below, I describe this process as an elite applying its power 'defensively' (Salverda 2010). Hence, as elsewhere, 'land remains a particularly vexing and contested form of property. It is both material and symbolic' (Fay and James 2010, 42).

At the same time, Franco-Mauritian landownership – and economic power more generally – forges collaboration between the government and the Franco-Mauritians. Investment of wealth originating from Franco-Mauritian landownership – and their control of the sugar industry – has contributed to the economic development of the island from the 1980s onwards. Recently, the conversion of agricultural lands was equally the result of a combined effort between landowners and the government to improve the economic conditions on the island. With the sugar and textile industry in decline, the search for new economic inroads has given Franco-Mauritian land new value. To attract foreign investment, the government initiated the Integrated Resort Scheme (IRS), that is the possibility for rich foreigners to buy residential property. Under this scheme, Franco-Mauritian landowners converted their agricultural land into luxurious villa resorts. In the postcolonial reality, then, Franco-Mauritians apply their land in different ways to sustain their economically powerful position. Hence, a closer look at the ambiguity of Franco-Mauritian landownership, which on the one hand instigates resentment but on the other shapes collaboration, helps to enhance the wider understanding of how land facilitates the 'defence' of an elite position. This helps to explain how landed (white) elites are, in the absence of (state) violence, able to maintain their positions long after the end of colonialism – yet, this equally explains why 'land-issues' will remain omnipresent in postcolonial Africa in the foreseeable future.

Unexplored territory

Mauritius was uninhabited when the first European seafarers set foot on the island in the sixteenth century. The French, who colonised Mauritius after the Dutch, started a permanent settlement in 1721, leading to the development of the island (Ly-Tio-Fane Pineo 1993). These settlers can be considered to be the ancestors of the present Franco-Mauritian community, although they were not a cohesive group from the start, as the island attracted whites of various social ranks during the French period (Vaughan 2005). The French East India Company tried to attract settlers by granting them land concessions. The absence of an indigenous population facilitated this and the first concession was granted on 5 June 1726, with successive concessions geared towards giving colonisation of the island an essential boost by distributing land to white settlers (Ly-Tio-Fane Pineo 1993, 98). Despite these efforts, the French ambition to develop the island into a plantation economy was barely achieved. Most of the large land concessions were only marginally exploited by 1766. At that time commerce seemed more profitable than agriculture. However, commercial trading weakened towards the end of the French period, and with few alternatives, the colonists had little choice but to turn their attention to further development of the sugar industry (Allen 1999, 21–22) – due to the hazards of cyclones in this part of the Indian Ocean region, sugarcane was considered the least risky crop to grow. Although expansion of the sugar plantations was encouraged further through access to the London market, which came with the inclusion of Mauritius in the British Empire, it was not until the late 1820s that sugar began to fully dominate the island's economy (Benedict 1965, 15).

In December 1810, during the Napoleonic war, the British arrived with an armada of 70 ships and forced the French out of Mauritius. The British took over the island to help safeguard their interests in India. They only took control of the island's administration and never made attempts to completely control the island or evict its French citizens. Property was not confiscated and most of the plantations remained in the hands of white French planters. Moreover, Britons never settled in large numbers and, consequently, direct British cultural influence was minor with French remaining the dominant European culture and language (Benedict 1965, 13–14). This, with the sugarcane plantation economy, also marked the gradual unification of the Franco-Mauritians as an elite, owning large tracts of land and distinct from the rest of the population by their white skin-colour and European ancestry (Salverda 2011).[4] As such, the small Franco-Mauritian community came to dominate the economic and political activity in Mauritius (Mozaffar 2005, 270).[5]

Shortly after their arrival, the British posed the first real challenge to the Franco-Mauritian elite position by abolishing slavery. This was part of an attempt, initiated in Britain, to ban slavery globally which was extended to Mauritius, rather than other Mauritians opposing Franco-Mauritian hegemony locally. The intention to abolish slavery was met with much resistance from the slave-masters, though they were unsuccessful in their attempt to stop the British from finally abolishing slavery in 1835. But they did manage to negotiate and obtain a huge compensation for the loss of their slaves. Franco-Mauritians did not have enough political clout to determine the decision-making process, but their economic position still gave them major influence in order to negotiate compensation. A side effect of the power struggle, moreover, was to the advantage of the Franco-Mauritians: they reinforced their elite position because they were unified not only by their shared economic interests but also by their joint resistance to British interference in their affairs (Vaughan 2005, 262). '[T]he Mauritian plantocracy as a whole, therefore, did not suffer drastic property losses from [slave] emancipation' (North-Coombes 2000, 23), and they remained powerful in the political and economic domain. But they had to turn to another source of labour. This was India, which they accessed with the help of the British colonial government. Thus began a new episode in Mauritian history, and the establishment of a new potential counter-force to white Mauritian hegemony: within 10 years of the arrival of the first indentured labourers one-third of the population were Indians, while by 1861 they numbered two-thirds of the population (Benedict 1965, 17).

Change

During the colonial period, Franco-Mauritians gained wealth and obtained and controlled the island's land relatively easily. Contrary to mainland Africa, they did not confront indigenous populations occupying the land, while communal lands, like elsewhere in the world, did not exist. From the onset, Mauritius, as a matter of fact, formed part of the capitalist world system, relying on the export of its sugar. This reliance on a mono-crop is also explanatory to the understanding of the Franco-Mauritian elite. In Nyasaland (present-day Malawi), for example, the white farming community faced conflicts of interests between producers of cotton, tobacco and tea, between different regions, and between well-endowed company estates and undercapitalised family farmers (Palmer 1985, 215). In Mauritius, the focus on one product united the landed elite, also called the sugar oligarchy, even though when the process of production from planting to milling and trading is divided among several groups, as in the case of the Philippines, there is a potential for clashing interests (Billig 2003). But the Franco-Mauritian sugar estates also controlled the sugar mills. Together with the British colonial government, they had a shared interest in the maintenance of a large and cheap labour force for the production of sugar (for the British market), which now were Indo-Mauritian descendants of Indian indentured labour. Any resistance was suppressed without much difficulty: the colonial system was a hegemonic one facilitating power for white Franco-Mauritians. Only a small group of relatively wealthy non-whites called *gens de couleur* (people of colour), descended mainly from manumitted slaves and (illicit) sexual liaisons between white and black people, had been granted equal rights and were allowed to participate in the island's affairs (Boudet 2004, 53). They were, however, a minor threat to Franco-Mauritian domination, because they were relatively few, less wealthy and continued to be excluded from Franco-Mauritian society (Allen 1999, 79–104). Nevertheless, changes crept in that paved the way for challenges to the Franco-Mauritian elite position. These were prompted initially more by developments in world markets in the late nineteenth century than they were by resistance to colonial rule: falling sugar prices led Franco-Mauritians to sell part of their land, predominantly to Indo-Mauritians, thereby allowing the latter to become gradually more prominent in the island's economic and political affairs.

The Franco-Mauritian estate-owners had problems in raising capital and by selling land they were able to extract substantial sums of ready cash from the Indian immigrants and their descendants – this process was referred to as *grand morcellement*. These Indo-Mauritians then became small landowners and planters themselves, which allowed them to steadily increase their economic and political power. South Africa, on the contrary, witnessed a reversed process with the white community appropriating 90% of the land surface under the 1913 Natives Land Act (Hall and Ntsebeza 2007, 3). As Allen argues, this access to and control of land was crucial to enhancing the standing of the Indo-Mauritians in the Mauritian colonial order (1999, 73–74, 138). But with support from the British, the Franco-Mauritians prevented any change that would seriously jeopardise their elite position, such as widening suffrage, for another half a century.

The British attitude, however, was about to change when the 1930s saw the emergence of serious challenges to the Franco-Mauritian position as large numbers of Mauritians who had been excluded from the island's affairs began to prosper. Observing the increasing influence of the European working classes, who had already proven a substantial challenge to the establishment (Hartmann 2007, 5–7), the Mauritian working classes, notwithstanding their ethnic background, started to raise their voice. This was fuelled by dissatisfaction with their situation and the global Great Depression of the 1930s that had devastated the economy of Mauritius, which was solely reliant on its sugar exports. The situation further deteriorated when the sugar mills reduced the buying price for a specific cane mainly produced by small planters. The sugar mills were in Franco-Mauritian hands, as was the control of the island's scientific institutions – 'access to sugar cane was a central grievance of protestors' (Storey 1997, 142). In August 1937, this led to a number of strikes and riots pitting workers against their employers in the sugar plantations. These were the first riots in Mauritian history which also had fatal casualties. These riots are considered a turning point in Mauritian history; the British colonial government could not ignore the grievances of the working classes anymore and as a result, changes were made: 'the Mauritian government began to incorporate non-elite groups within the structures of the state, to guarantee the peace' (149).

After the Second World War pressure built on the Franco-Mauritian position, as the general shape of events was influenced by new anti-colonial superpowers, the birth of the United Nations and a significant shift in the metropolitan view of empire (Jackson 2001, 152–155). Until then there had been mutual support and close cooperation between the British and Franco-Mauritians. But under growing pressure from non-white Mauritians, especially Indo-Mauritians, the British drafted a new constitution in 1947, which widened suffrage substantially. Franco-Mauritians were furious with the British because the colonial administration virtually ignored their suggestions for the new constitution. This marked a change in the relationship between the two parties. The Franco-Mauritians knew that a radical change in suffrage would increase the political power of the Indo-Mauritians, especially the majority Hindus. But the British had already decided that, for the well-being of the colony, the labouring classes and the upcoming (counter-)elites should be given a voice. This change in attitude, which led to the extension of the franchise to all adult literate men and women in 1956 (Seekings 2011, 172) and was a clear defeat for the Franco-Mauritians, had Mauritian independence in 1968 as its apogee.

Franco-Mauritians lost their political dominance in the transition (Salverda 2010, 394–396; Seekings 2011, 172), which led many to rethink their position. They felt anxious about their future and the island's economic prospects under Hindu rule and a number of them left for South Africa (Boudet 2004) despite the fact that their economic base was scarcely challenged. Contrary to Zimbabwe, for example, where land redistribution featured prominently on the agenda at the time of independence, landownership was less of an issue in Mauritius. Franco-Mauritian economic dominance could be maintained because the new political elite, predominantly of Hindu background, initially sought political power only and were able to identify areas of common interest with the Franco-Mauritians. As Simmons noted, 'both understand that political instability would destroy their own positions' (1982, 193). Also, South Africa witnessed no large-scale asset redistribution in the transition, while, contrary to the prominence of land redistribution in the prelude to independence, the issue of land reform fell silent during the first decade of Zimbabwean independence. The new regime hardly challenged the white commercial farmers in the 1980s. At the time of independence, partly as a result of the war, white commercial farmers were producing some 90% of the country's marketed food requirements. According to Palmer, Mozambican's Frelimo, the new government's wartime ally, advised Zimbabwe's new rulers to retain white experience and skills. Frelimo had cared little about the exodus of Portuguese settlers when it fought its way to power in 1974, but later came to regret this (Palmer 1990, 167). The whites were crucial to the economy. Zimbabwe, as history tells us, has since changed its approach on white landownership. Mauritius has not followed the same path, yet the colonial legacy of landownership, as the next section will illustrate, is both a blessing and a threat for the Franco-Mauritians.

The ambiguity of landownership

Today, Mauritius is classified as a middle-income country due to economic prosperity from the mid-1980s onwards, which was the result of the diversification of the economy away from an economy solely relying on the mono-crop, sugar. This had a positive impact not only on the population in general, but also on the Franco-Mauritians who were largely involved in the diversification. Sandbrook et al. argue that, '[r]evenues from sugar exports were used to diversify and industrialize the economy, as well as support an expanding welfare state' (2007, 124–125). In the 1970s, the establishment of the export processing zone (EPZ) took on form, an export-oriented manufacturing venture best known for textile production which, among other things, benefited from special investment promotion incentives and exemptions from customs duties (Bräutigam 2003, 458; Neveling 2006, 2). According to Handley (2008, 109), '[b]etween 1970 and 1983 Mauritian businesspeople provided almost half the total equity capital to the new sector – and 47% of that local equity capital came directly from investment by private sugar companies'. A renowned Franco-Mauritian businessman from a well-established family that was involved in the sugar industry for generations explained to me how his family, in the early 1970s, initiated activities in the EPZ – with the assistance of the Mauritian government, as the Mauritian Ministry of Industry, Commerce and International Trade brought him into contact with a European agency specialising in connecting investors and producers. Around 1990, his family ceased their sugar activities altogether and sold most of their land, and chose to focus on a variety of ventures in the EPZ as well as on other businesses in which they own shares (interview with author). The export-oriented nature of the EPZ, however, has also created opportunities for other Mauritian businessmen. They did not need large plots of land for constructing factories, while they could also avoid the Franco-Mauritian dominated local market since they mainly dealt with international clients. The EPZ, moreover, influenced the social and economic composition of Mauritian society as a whole. In 1986, the sum of EPZ industries became the largest employer in Mauritius, surpassing the sugar industry. This led to economic prosperity for many of the island's inhabitants and to a high rate of social change. Unemployment decreased significantly, women increasingly participated in the labour market and the large EPZ factories were more ethnically diverse than the traditional businesses, thus leading to increasing interethnic contact (Eriksen 1998, 153–157).

Wealth originating from the sugar industry was also invested heavily in tourism, thereby contributing to the establishment of Mauritius as a luxurious holiday and upmarket honeymoon destination. The direct impact of landownership was more decisive in this venture than in the EPZ. Franco-Mauritians may not have owned the beachfront, as all land directly bordering the sea is owned by the state, yet they did often own the land behind it. This was an advantage, as they could use this land for constructing hotels and golf courses. Apart from the large international hotel chains, then, the luxurious segment of the market is dominated by Franco-Mauritian controlled businesses. According to Handley, '[tourism] was one of the few areas where the private sector in Mauritius [in this case the Franco-Mauritians] played the lead role in developing a vision for its long-term development. (It was only as late as 1988 that the government established a specific ministry for tourism.)' (2008, 111–112). From that period onwards tourism has contributed to the prosperity of many Mauritians, but it has also largely consolidated Franco-Mauritian economic power.

Evidently, Franco-Mauritian landownership has contributed to the consolidation of economic power. In 2007, of the 10 largest Mauritian business groups (based on annual turnover) five were Franco-Mauritian and they all were or were linked to large landowners – either the business group has a branch managing land and sugar industry assets or substantial share ownership is related to families whose wealth originates from the sugar industry. These business groups may have interests in tourism and/or other sectors, but this cannot be disentangled from their land possessions. A foreign businessman with a long experience in the Mauritian private sector, consequently remarked, 'without [control of the] land there would be no Franco-Mauritians, because the land is the wealth they rely on' (interview with author). How much of the island's land the Franco-Mauritians possess at the moment is difficult to establish. A general estimation suggests that Franco-Mauritians own approximately 36% of the total available land, predominantly agricultural land (L'Express Dimanche, 13 May 2007; L'Express, 31 May 2007). Only about 10% of the island's land is state-owned (L'Express, 31 May 2007), while another portion of the total land is made up of small plots of residential land (L'Express Dimanche, 13 May 2007). The community's wealth may no longer be as strongly linked to sugarcane as it once was; the colonial origin of the sugarcane industry and concentrated land possessions remain central to the Franco-Mauritians' relatively successful maintenance of an elite position more than 40 years after independence. In the light of this historical attachment to the sugar industry, Franco-Mauritian identity, Boudet argues, is not only marked by their white skin-colour, the French language and 'French' culture, but also by their monopoly within the sugar industry (2005, 24) – even for the Franco-Mauritians who were or are not directly involved in the industry. In other (former) colonies, one can observe similarities to this situation. For example, in Martinique, where the sugar industry has substantially declined in economic importance, 'the plantation past represents a kind of "golden age" for Békés [the island's white elite]' (Vogt 2005, 88). One Franco-Mauritian argued that possessing a mill is a significant status symbol among economically powerful Franco-Mauritian families (interview with author) – this relates to the possession of land, because in the Mauritian case the largest landowners also control the mills.

Over time, the business groups and families owning land have decreased due to takeovers and mergers. Landownership has become much more concentrated, with a small number of business groups and related families owning most of the land. In virtually all cases, ownership is not in the hands of single individuals but in the hands of family holdings that can have up to a hundred shareholders. A businessman involved in the management of the wealth of one of the island's economically most powerful family holdings, with 80 to 100 shareholders, explained that it is, 'not that all the heirs of my great-grand father, who started the small sugar estate in 1912, are equally wealthy. Some have made their own investments or sold out their shares to other family members' (interview with author). In other cases, ownership is divided among different Franco-Mauritian businesses and/or family holdings, this also explaining the relatively tight Franco-Mauritian business networks. In cases where Franco-Mauritians cannot gather sufficient capital, they often turn to other Franco-Mauritians to share the investment. This also explains the wide portfolio of the large Franco-Mauritian business groups, since these are often the result of takeovers and mergers of (family) businesses. This is not to say that there is no animosity and infighting between Franco-Mauritian businessmen and business families - or even within families. But through the Franco-Mauritian network, businesses in financial difficulties and/or with succession problems have always been first offered to other Franco-Mauritians. The successful consolidation of economic power and land-ownership is partly the result of this historical pattern: wealth may have changed hands between Franco-Mauritian families but it remained within the Franco-Mauritian community. Besides, conflicts have predominantly been about control and/or strategy and not, as in other cases, between different sectors of the economy. Franco-Mauritian landowners have predominantly applied their lands for the same purposes, be it sugarcane production, tourism or today's IRS schemes. Hence, notwithstanding some diversity and internal competition among Franco-Mauritians, they tend to have a rather uniform economic vision. As a Franco-Mauritian CEO of a family holding with large interests in the sugar industry said, '[t]here is no hardship between the different [Franco-Mauritian] groups. We compete, but not until the bitter end because socially we are also friends. I consider the competition more as a game with sportish rules' (interview with author). Understandably then, also because of its association with colonial inequalities, the correlation between Franco-Mauritian economic power and landownership remains a prominent issue. Unlike elsewhere in Africa, Mauritius may have never had large-scale policies aiming to (re)distribute land, yet the issue of landownership is prominent. According to one Mauritian journalist, 'the unequal distribution of land is at the centre of the problem; without a change nothing will happen' (interview with author).

Pay-off

The Franco-Mauritian relatively successful maintenance of economic power well into the twenty-first century is a recipe for resentment. Politicians openly criticise this power by, for example, using strong symbolic rhetoric about the sugar or even white oligarchy. In other cases, they just simply refer to the five (Franco-Mauritian) families – as the supposedly most powerful families economically. In 2005, a new government was elected after a campaign that had been rampant with 'white-bashing' that is (indirectly) criticising the economic power and privileges of the Franco-Mauritians.[6] This is not unusual during election campaigns, and Franco-Mauritians are aware of their role as targets – they tend to ignore the criticism voiced during election campaigns and present an image of neutrality. Franco-Mauritians argue that politicians, in order to gain votes, criticise Franco-Mauritian economic power because, as a Franco-Mauritian CEO said, 'there are so few whites that if the [political] mechanism of "white-bashing" doesn't work for you, it doesn't work against you' (interview with author). A widely shared perception is that after the elections politicians tend to tone down their criticism because in the end the private sector and the government need each other – a Franco-Mauritian businessman said, 'when I'm having a drink with politicians they tell me that [white bashing] was just talking politics' (interview with author). This ambiguous relationship between public rhetoric and private consent to the status quo has gradually led to consensus among Franco-Mauritian businessmen that it is best to support the government in place and remain neutral during the electoral campaign. In the government's criticism of Franco-Mauritian unequal economic power, moreover, landownership often features prominently, notwithstanding that unlike other countries on the African continent, Mauritius does not have a large class of peasants or groups that previously occupied the land. The close relation between landownership and economic power, however, seems to justify the aim to distribute the land more fairly. Paradoxically, political actions geared towards offsetting the unequal possession of land give land a prominent function in negotiations with the government. My concept of 'defensive power' (Salverda 2010) is key to understanding this.

Elites are often considered as all-powerful because of the view that elites, through their control over resources (such as land), have the most power at their disposal. The assumption is that they are the ones exercising power proactively and expansively. But it needs to be stressed that elites are not necessarily all-powerful (Scott 2008, 38), and that they, especially in the face of change, tend to defend their interests and privileges as a reaction to external challenges to their position. In the transition to independence, Franco-Mauritians, for example, reacted to pressure from especially the Hindu population of Mauritius that aspired to obtain a political say in the affairs of the island. Elites, then, apply their power to resist pressure in order to maintain the status quo. History indicates that elites have had varying success in defending their interests and privileges. It is obvious that colonial and white elites have lost their hegemony – their initial dominance over virtually all (public) spheres of life. Their relative power has declined, though not necessarily their absolute wealth. In a number of cases, like South Africa and Mauritius, white elites have succeeded relatively well in maintaining an economically powerful position. This shows that elites have to move from exercising power over others directly to applying their (remaining) resources in such a way as to prevent them from losing their power base and privileges. In Weberian terms of the workings of power, the elite does things it would not otherwise have done due to the exercise of power by others. One could argue then that the elite resists, as if they were subalterns, yet I argue that from an analytical perspective a distinction ought to be made between an elite applying power defensively and identifying an action as subaltern resistance. As an analytical concept, the resistance of subalterns should be considered as the means to try to undo an unbalanced situation. The two principal forms of subaltern resistance - pressure and protest - are active forms employed to challenge the established power structure (Scott 2001, 27). The elite, however, applies power defensively in order to achieve maintenance of the status quo instead of trying to alter the situation. Thus, in this case the elite is more passive, instead of proactively using its power. Demands for the redistribution of land in Mauritius aptly illustrate this.

In Mauritius, sugar traditionally has been the 'country's cash cow' (Handley 2008, 108–109), not only as an important source of revenue, but also through a pattern of applying economic power in the form of financial contributions and donations paid by Franco-Mauritian businesses to government-related projects. According to a retired Mauritian businessman of Muslim origin, in the deals between the government and Franco-Mauritians land is often exchanged (interview with author). For example, the Mahatma Gandhi Institute (MGI), which promotes (research on) Indo-Mauritian culture and was founded by the Mauritian government in collaboration with the Indian government in 1970, is situated on a plot of land donated to the government by a large Franco-Mauritian business group. Here land helped to smooth relations with the government. This pattern of obtaining support through the distribution of land actually has a long tradition, as many Franco-Mauritian sugar estates give – or sell cheaply – plots of land to civic and religious groups, which are often affiliated with their workforce. Franco-Mauritians hereby have to seek a balance between maintaining control over resources and smoothening relationships with larger social groups or the government – the latter, of course, serves the former. One sugar estate, for instance, wanted to change its policy of parcelling out land in a haphazard manner. Instead it decided to draft an integrated plan that set out the development of the area. One of the shareholders in the estate said, 'not all the directors of [the sugar estate] found the [new] ideas convincing, because too much was given to the public and the government. [But] times have changed and this is the way to do it. It was really well received by the government' (interview with author).

Land also proved key to suppress tensions between the Franco-Mauritian sugar plantations and the government when the sugar industry plunged into recession in 2005. Reform – which led to tense negotiations with the government – was needed to alter the dire state of the sugar industry. In the tense atmosphere, land redistribution became an issue, showing once more that land is not only the cause of numerous power struggles between the government and the Franco-Mauritian private sector but also a potential means to ease the tensions. Landownership put the Franco-Mauritian-controlled sugar estates in a precarious political position, despite the fact that, as said, the Mauritian economy nowadays relies far less on the sugar industry than before. At the time of the negotiations, agriculture, of which around half of the added value is sugarcane-related, made up only 6% of Mauritian GDP.[7] In densely populated Mauritius, the land associated with the sugar industry, however, is highly symbolic and reinforces resentment even though the distribution of land is a complex matter in Mauritius because initially the land belonged to no one. To a certain extent, then, the Mauritian case is different to most cases on the African continent. Zimbabwe, for example, witnessed a violent redistribution campaign of land occupied by white farmers – the land has been expropriated on the basis that it had been unfairly appropriated by white farmers. This also shows the limits of defensive power once a powerful opponent turns to violence. Initially, there was a certain level of peaceful land distribution (Palmer 1990; Shaw 2003, 75–76), but once Robert Mugabe launched his aggressive land distribution campaign, there was little the white minority could do.

Reform would involve the closing of sugar mills and, subsequently, social programmes for laid-off workers – in these cases the government often demands land from the sugar industry to bring these social programmes to a successful conclusion. Initially, a deal was struck between the sugar estates, the government and the European Commission (EC) (which was willing to contribute financially to the reform). But the government stalled and brought the issue back to the negotiation table, demanding extra compensation of two thousand *arpents* (one *arpent*, an old French unit for measuring land, is about half a hectare) to be paid by the sugar industry for social projects, as it considered the deal to be too advantageous for the Franco-Mauritian sugar industry. The result was a deadlock. In this conflict, Franco-Mauritians perceived their position to be openly under threat and felt that they had to stand up for their rights – contrary to them normally keeping a low profile and presenting an image of neutrality. Now they found themselves unable to avoid direct confrontation. A Franco-Mauritian businessman, without interests in the sugar industry, said, '[the sugar estates] don't know what's next; they wonder what happens if they give in, "will the government then come with other demands?"' (interview with author). The Franco-Mauritian sugar estates accused the government of making excessive demands and not respecting the rules of fair play; at no point had the extra compensation been brought up in the (initial) deal, they argued.

Following up on Boudet's (2005) analysis of Franco-Mauritian identity, even Franco-Mauritians without landed interests supported the sugar estates. A Franco-Mauritian couple argued that the politicians suddenly wanted land to distribute among their political friends. The distribution is not in favour of the poor, they stated (interview with author). This support for the landowners also needs to be examined in the context of the 2005 political campaign. Franco-Mauritians felt victimised and realised that politicians did not stop criticising Franco-Mauritian economic power and privileges after the election campaign. The government proposed, for example, to 'democratise' the economy, that is a more equal distribution of economic gains nationally. Franco-Mauritians felt that this was directly targeted at them. Echoing the thoughts of the Franco-Mauritian couple, many Franco-Mauritians argued, in private, that there was a further hidden agenda to the democratisation of the economy, namely consolidation of the prime minister's personal power. According to this view, his intention was to take the wealth from the whites, in order to distribute it to his own community and other proxies. Also, in 2006 the government decided to change the conditions for the lease of the *campement* (i.e. seaside bungalow) sites. The *campements* and seaside life are a very significant element of Franco-Mauritian elite culture, and Franco-Mauritians considered the new policy to be a threat to their lifestyle. Franco-Mauritians argued that the increase in the lease price was exorbitant and that the government proposal was targeting them as whites.

In the negotiations about the reform of the sugar industry, the Franco-Mauritians defended themselves, but in the end they had to give in to government pressure. Then the government came back with yet additional demands. Again, the sugar industry said that it could not possibly meet these demands, before eventually agreeing to satisfy a substantial part of them. The two sides subsequently came to an agreement. The final result: the sugar industry gave two thousand *arpents* of land for social programmes and opened up 35% of the shareholding of the mills (*L'Express*, 6 December 2007). It was obvious that the Franco-Mauritians and the government were highly dependent on each other and needed to come to an agreement in order to safeguard the European Commission's financial contribution to the restructuring programme since the EC demanded that the Franco-Mauritian sugar industry and the government come to an agreement (*L'Express*, 15 May 2007). Arguably, the Franco-Mauritians conceded the most, but not all the government's demands were met. In essence, Franco-Mauritian economic power could not compete with the mobilisation of political power by the government, though it contributed to manoeuvring them into a negotiating position.

The Franco-Mauritian elite position was hardly jeopardised, because it is especially the balance between opposing and giving in that explains the success of defensive power. Franco-Mauritians often feel victimised and oppose change that may jeopardise their position. In the case of the democratisation of the economy, for example, many Franco-Mauritians said that, in principle, they supported the idea of sharing the cake with everyone but opposed the idea of taking wealth from one person (i.e. Franco-Mauritians) and giving it to another (i.e. the prime minister's cronies and supporters). With respect to this issue, repeated comparisons have been made with Zimbabwe where Robert Mugabe expropriated white farmers' land, causing the free fall of the economy. A Franco-Mauritian politician justified this comparison by arguing that the government wanted to throw out the whites. Paradoxically, to fully support the Franco-Mauritian case he also needed to downplay the comparison. 'The situation is different than in Zimbabwe, because [the landowners] haven't taken anyone's land', he said (interview with author). It is hard to know if this comparison constitutes a strategy to instil anger against the government or a real fear. The rhetorical comparison with Zimbabwe certainly helped to label government's proposals as anti-white, and as a consequence, the few 'dissident' voices in the Franco-Mauritanian community who have pointed out the merit of some of the proposals are hardly heard. Franco-Mauritians claim that they are in favour of true democratisation of the economy, but by discrediting, rightly or not, the government's intentions, they end up simply resisting, without contributing to, this process. But as the redistribution of land shows, Franco-Mauritians cannot oppose only. They have too many vested interests to completely alienate themselves from the government and thus have to give in to demands to safeguard their position – this may be different from white farmers who solely rely on their farming activities. Ceding resources (land), after all, helped to appease the government. The success of defensive power, thus, relies on the potential to partially handover (the control over) resources, be these land or other resources. In this sense, the Creole elite in Sierra Leone (Cohen 1981) and the Parsi elite in India (Luhrmann 1996) were at a disadvantage, as they controlled few tangible resources that they could apply to defend their position. Success, however, seems to also rely on the elite's opposition, as in the case of the Franco-Mauritians, their loud opposition to government demands manoeuvred them into a negotiating position. Now they could prevent more substantial distribution of resources. I doubt, however, whether this was strategically intended. It seemed more the result of the evolution and solving of the dispute with the government. The landowners' stand also appeared to be influenced by the fact that not only does the land for them have an economic function, but also they are emotionally and symbolically attached to the land. For them, opposing land redistribution was not purely driven by economic interests. Nevertheless, finding a balance between opposing change and accepting a partial distribution of land gives Franco-Mauritians respite and helps them to prolong their elite position, even though this means a gradual decline of their landownership. At the same time, ownership of land is functional in collaboration with the government, and as such also contributes to the maintenance of their elite position.

Collaboration

As already illustrated, the Franco-Mauritians in the private sector and the government have incentives to collaborate. Both have interests in the economic prosperity of the island, which to a certain extent protects Franco-Mauritian private property. The Mauritian government adheres to a global political-economy favouring private property. This has always been the case, because, as said, its origins lie within the global capitalist system. Besides, Mauritius does not have a large peasant population, neither have populations been displaced, nor does it have a tradition of communal lands. Mauritius being a democracy the constitution also guarantees the protection of deprivation of property. Notwithstanding resentment, then, both sides share similar economic views. The government does not want to jeopardise the future of the island by completely disregarding Franco-Mauritian ownership. The same seems to apply for land redistribution in South Africa even though it is of a completely different character. Here the government equally appears to balance between a strong push for redistribution and a conviction that this push may jeopardise the economic future.

Handley argues that the private sector 'tended to follow the policy lead given by the government rather than proactively initiating policy change' (2008, 116) – the exception, as I have cited above, is the tourism industry. Government involvement, for example, has been a windfall for the Franco-Mauritian sugar industry. There has always been close co-operation between the Franco-Mauritians and the government regarding the issue of access to the European market under advantageous conditions, because the sugar industry has been an important employer and revenue provider (Hein 1996, 23). The government has constantly been a strong negotiator in dealings with the European Commission for selling Mauritian sugar to the unified European market. Through the Sugar Protocol of the Lomé Convention(s) and the Cotonou Agreement, Mauritius (under the umbrella of the ACP (African, Caribbean and Pacific) countries) negotiated quotas and preferential prices for their sugar.[8] These prices were well above world market prices. Considering the fact that Mauritius was the biggest beneficiary of these negotiations (Hein 1996), the Franco-Mauritians, who firmly controlled the land and the mills, thus gained substantial profits. Moreover, in spite of tense negotiations, collaboration was equally witnessed at the time of the 2005 recession in the sugar industry. With the textile industry also in decline, the government and the private sector actively tried to search for new economic inroads, thus giving Franco-Mauritian land new value.

One of the means to boost the local economy was to attract foreign investment: the government initiated the IRS, that is the possibility for rich foreigners to buy residential property. Under this scheme, Franco-Mauritian landowners converted their agricultural land into luxurious villa resorts, to allow foreign nationals to reside in Mauritius by investing a minimum of US$500,000 in local property (Kothari and Wilkinson 2011, 11). Numerous such resorts have been built around the island, on converted sugarcane fields, and often overlooking the Indian Ocean. The colonial heritage of Franco-Mauritian landownership proves to be a substantial asset, as the IRS is a massive advantage to those possessing prime land. Others would have to buy this land at market price, which makes it difficult to compete. In many of the projects, Franco-Mauritians collaborate with foreign investors: they provide the land, the foreign companies the financial means, knowledge and/or international networks. As a CEO of a smaller and independent sugar estate said, 'Franco-Mauritians are assets rich because of their land, but cash flow poor' (interview with author). Hence, Franco-Mauritian landowners substantially profit from this scheme, even though it is in close collaboration with the government – which hopes that it will increase employment as well as revenue. Their move to sell off land, however, is perceived as unusual. According to one Franco-Mauritian, the IRS scheme is a break with the past, as Franco-Mauritians are now for the first time selling their land (interview with author) – this, as the case of the *grand morcellement* shows, is not historically accurate, though it reveals the symbolic association of Franco-Mauritians with landownership. In terms of smooth relationships with the government, Franco-Mauritians need such collaboration, though it certainly brings them income as well. Economic development may benefit the whole of the country – Mbeki (2009) actually considers Mauritius as a successful example of an inclusive development in sub-Saharan Africa – it certainly enhances the wealth of landowners. The role of attachment to the (is)land should not be disregarded here, as Franco-Mauritians – but it also seems whites in, for example, South Africa – do reinvest in the country instead of only extracting wealth.

Conclusion

The Franco-Mauritian case has its unique features, especially because Mauritius was uninhabited before the settlement of European colonisers. This case, however, gives relevant insights into how landed (white) elites are able to maintain their positions long after the end of colonialism. It offers theoretical insights into how control over land is simultaneously a disadvantage as well as an advantage when (landed) elites are faced with challenges to their position. Franco-Mauritians continue to face political pressure as a result of their association with their colonial past and their control over much of the island's agricultural lands. This has forced them to actually cede parts of their lands. However, landownership also supports Franco-Mauritians facing challenges of decline, in the sense that its redistribution offers some remission from both political and economic challenges. The result is a balancing act, as Franco-Mauritians defend themselves against too much loss of power (and land) at the same time as they also have to give in to some of their opponents' demands. It is to be noted that this concerns economic power more generally, though land is an important foundation of resentment and the means of 'buying time' to slow the complete redistribution of this resource. This pattern of balancing act can equally be observed in South Africa:

> in the relationship between business and government, the government did not look kindly on critical public remarks by prominent businesspeople. On many issues then, organized business chose to cooperate with government and express its concerns behind closed doors. This may have contributed to some degree of self-censorship. (Handley 2008, 96)

This refutes Mbeki's claim that 'the political elite is comparatively more vulnerable since the economic oligarchy can promote new political parties to challenge a hostile ruling party' (Mbeki 2009, 83). Both cases actually illustrate that this would be highly risky to do. It is predominantly the approach of balancing between giving in and a certain level of opposing change that contributes to the maintenance of economic power – respectively, more than 40 years after independence and almost 20 years after the end of apartheid. This, however, seems to be successful only in the absence of (state) violence – Zimbabwe constituting the obvious example of the limits of defensive power.

The Franco-Mauritian case shows that opposing sides in a peaceful country characterised by compromise can simultaneously collaborate, with the potential of giving new value to landownership at the same time as the economy is stimulated and new employment provided. For fear of losing revenue – from tourism as well as from other sources – the Mauritian government does not want to jeopardise the image of Mauritius as one of Africa's the most stable states. The government, then, is required to collaborate with the Franco-Mauritians in the private sector. This, consequently, reinforces the Franco-Mauritian elite position, yet at the same time also lays new foundations for the resurfacing of resentment about the unequal distribution of resources. With all the associations with colonial injustices, this resentment may easily resurface at a later moment in time, during election campaigns for example. The need to collaborate may partly take the pressure off the Franco-Mauritians, yet public opposition is likely to return. The South African case, though substantially different, is not likely to be solved quickly or neatly either. Resentment continues to run high, forcing landowners to redistribute all or part of their landholdings. Their opposition, however, will slow the complete redistribution of their land, while at the same time the government may not want to push it too far out of fear of jeopardising food security and the South African economy more generally. While the Franco-Mauritian case is unique inasmuch as Mauritius was uninhabited before colonisation, it does reveal the ambiguity of landownership in postcolonial settings, and explains why 'land-issues' will remain omnipresent in postcolonial Africa for the foreseeable future.

Notes

1. This article is based on historical data and ethnographic fieldwork conducted in Mauritius, South Africa and France in the period 2005–2007. Over 125 interviews were conducted with Franco-Mauritians in Mauritius, South Africa and France, while 30 official interviews were conducted with other Mauritians. The research also relies on informal conversations, participant observation, a questionnaire I conducted, and media research, both archival and from contemporary sources.
2. The total population of Mauritius is about 1.3 million. Apart from Europe, Mauritians have their origins in distant locations such as China, India and Africa. The largest group are Hindus (52%) and there is a minority group of Muslims (16%). Both groups originate in India and most came as indentured labourers when slavery was abolished in 1835. Creoles, largely of slave descent, constitute about 27% (including a small group of gens de couleur). Then there are the Sino-Mauritians, who make up 3% of the population (Eriksen 1998, 15).
3. Notwithstanding that Mauritius did not witness the displacement of people like elsewhere in Africa, there are certainly cases of Franco-Mauritians applying cunning ways to repossess the land of other Mauritians. A shared Franco-Mauritian background was not necessarily an insurance to be safe, though, because there are also stories of disputed appropriation among Franco-Mauritians – and even within families. One of the main recommendations of the Truth and Justice Commission (2011), in this light, is to set up a land monitoring and research unit in order to investigate and settle (historical) disputes regarding land transactions and title deeds.
4. The black population the Franco-Mauritians wanted to distinguish themselves from was a mixed group of slaves and free(d) persons originating from numerous locations in, predominantly, Africa and India.
5. Pointing to the remarkable life story of Marie Rozette – a freedwoman of Indian origin – who accumulated wealth, land and slaves in Mauritius in the late 1700s, during the French period, Allen (2011) observes this dominance was not comprehensive – equally there are examples during the British period of non-whites owning large tracts of land and slaves.
6. The role of the then prime minister, the Franco-Mauritian Paul Bérenger, was especially illustrative to this. Initially, Bérenger was not associated with the Franco-Mauritians, because when he started his political career in the first decade after independence, he strongly criticised Franco-Mauritian domination in the private sector. This helped him to gain wide support among Mauritians of all backgrounds. In 2003, Bérenger became the

first non-Hindu prime minister of Mauritius due to a pre-electoral agreement between his party and the party of Anerood Jugnauth, the prime minister who preceded Bérenger. That a non-Hindu became prime minister was thought to represent a break with the past and it was considered a sign that Mauritius was ready for decreased influence of ethnicity on politics. However, once in government Bérenger found himself obliged to cooperate with Franco-Mauritian businessmen, which made him an easy prey for political opponents. Now his skin-colour suddenly became 'visible': he was a 'white' favouring other 'whites'. His background clearly constrained him in dealing with private sector matters. In reality, Bérenger does not appear to have favoured the Franco-Mauritians or showed any 'racial' preferences. Yet, his white skin-colour was a liability. As a consequence of Bérenger's position as prime minister, in the ensuing 2005 election campaign there was a strong focus on the Franco-Mauritians. As many ordinary Mauritians are of the cynical viewpoint that government, in general, represents the interests of the Franco-Mauritians (Hempel 2009, 468), they easily accepted the political rhetoric that Franco-Mauritians hold all the economic power and, through the political figure of Bérenger, were becoming hegemonic again. This put the Franco-Mauritian community in a position of blame. Subsequently, Bérenger lost the 2005 elections; although it may be too simplistic to say that Bérenger cum suis only lost because of his white skin-colour.

7. See: http://www.gov.mu/portal/goc/cso/report/natacc/agri06/sumtab.pdf (accessed 7 December 2009).

8. http://www.acpsugar.org/Sugar%20Protocol.html (accessed: 7 December 2007).

Note on contributor

Tijo Salverda is a postdoctoral fellow with the University of Pretoria's Human Economy Programme. His interests lie especially in the study of elites and power. His PhD research (2010, VU University, Amsterdam) focused on how Franco-Mauritians, the white elite of Mauritius, balance continuity and creeping decline of their elite position. Tijo's current research interests are the investment side in large-scale land acquisition, and globally operating financial elites. His publications include 'Elite power: in defence', *Journal of Political Power*, 2010; 'Embodied signs of elite distinction: Franco-Mauritians' white skin-colour in the face of change', *Comparative Sociology*, 2011 and *The Anthropology of Elites: Power, Culture and the Complexities of Distinction*, Abbink, Jon and Tijo Salverda (eds), 2013, New York: Palgrave Macmillan. He can be contacted at tijo.salverda@up.ac.za or tijo@tijosalverda.nl

References

Allen, R. B. 1999. *Slaves, Freedmen, and Indentured Laborers in Colonial Mauritius.* Cambridge: Cambridge University Press.

Allen, R. B. 2011. "Marie Rozette and Her World: Class, Ethnicity, Gender, and Race in Late Eighteenth- and Early Nineteenth-Century Mauritius." *Journal of Social History* 45 (2): 345–365. doi:10.1093/jsh/shr058.

Benedict, B. 1965. *Mauritius: The Problems of a Plural Society.* London: Pall Mall Press.

Billig, M. S. 2003. *Barons, Brokers, and Buyers: The Institutions and Cultures of Philippine Sugar.* Honolulu, HI: University of Hawaii Press.

Boudet, C. 2004. "Les Franco-Mauriciens entre Maurice et l'Afrique du Sud: Identite, Strategies Migratoires et Processus de Recommunautarisation [Franco-Mauritians, Between Mauritius and South Africa: Identity, Migration Strategies, and Processes of Regrouping]." PhD diss., Université Montesquie, Bordeaux IV.

Boudet, C. 2005. "La construction politique d'une identité Franco-Mauricienne (1810–1968): le discours identitaire comme gestion de la contradiction [The Construction of Franco-Mauritian Political Identity (1810–1968): Identity Discourse as Means of Managing Contradictions]." *Kabaro, Revue Internationale des Sciences de l'Homme et des Sociétés* 3: 23–44.

Bräutigam, D. 2003. "Close Encounters: Chinese Business Networks as Industrial Catalysts in Sub-Saharan Africa." *African Affairs* 102 (408): 447–467. doi:10.1093/oxfordjournals.afraf.a138824.

Cohen, A. 1981. *The Politics of Elite Culture: Explorations in the Dramaturgy of Power in a Modern African Society*. Berkeley, CA: University of California Press.

Eriksen, T. H. 1998. *Common Denominators: Ethnicity, Nation-Building and Compromise in Mauritius*. Oxford: Berg.

Fay, D., and D. James. 2010. "Comparative Issues in the Study of Land Restitution." In *Land, Memory, Reconstruction, and Justice: Perspectives on Land Claims in South Africa*, edited by C. Walker, A. Bohlin, R. Hall, and T. Kepe, 41–60. Athens, OH: Ohio University Press.

Hall, R., and L. Ntsebeza. 2007. "Introduction to The Land Question in South Africa." In *The Land Question in South Africa: The Challenge of Transformation and Redistribution*, edited by L. Ntsebeza and R. Hall, 1–24. Cape Town: HSRC Press.

Handley, A. 2008. *Business and the State in Africa: Economic Policy-Making in the Neoliberal Era*. Cambridge: Cambridge University Press. doi:10.1017/CBO9780511491832.

Hartmann, M. 2007. *The Sociology of Elites*. London: Routledge.

Hein, P. 1996. *L'Economie de l'Ile Maurice* [The Economy of the Isle Mauritius]. Paris: L'Harmattan.

Hempel, L. M. 2009. "Power, Wealth and Common Identity: Access to Resources and Ethnic Identification in a Plural Society." *Ethnic and Racial Studies* 32 (3): 460–489. doi:10.1080/01419870701722422.

Jackson, A. 2001. *War and Empire in Mauritius and the Indian Ocean*. Basingstoke: Palgrave in association with Kings College London.

Kothari, U., and R. Wilkinson. 2011. "Global Change, Small Island State Response: Restructuring and Uncertainty in Mauritius and Seychelles." *Journal of International Development*. doi:10.1057/9781403919540.

Luhrmann, T. M. 1996. *The Good Parsi: The Fate of a Colonial Elite in a Postcolonial Society*. Cambridge, MA: Harvard University Press.

Ly-Tio-Fane Pineo, H. 1993. *Ile de France, 1715–1746* [Isle of France, 1715–1746]. Mauritius: Mahatma Gandhi Institute.

Mbeki, M. 2009. *Architects of Poverty: Why African Capitalism Needs Changing*. Johannesburg: Picador Africa.

Mozaffar, S. 2005. "Negotiating Independence in Mauritius." *International Negotiation* 10: 263–291. doi:10.1163/1571806054740976.

Neveling, P. 2006. "Spirits of Capitalism and the De-Alienation of Workers: A Historical Perspective on the Mauritian Garment Industry." Online Working Papers GSAA (2). http://www.gsaa.uni-halle.de.

North-Coombes, M. D. 2000. *Studies in the Political Economy of Mauritius*. Mauritius: Mahatma Gandhi Institute.

Palmer, R. 1985. "White Farmers in Malawi: Before and after the Depression." *African Affairs* 84 (335): 211–245.

Palmer, R. 1990. "Land Reform in Zimbabwe, 1980–1990." *African Affairs* 89 (355): 163–181.

Salverda, T. 2010. "Elite Power: In Defence." *Journal of Political Power* 3 (3): 385–404. doi:10.1080/17540291.2010.525829.

Salverda, T. 2011. "Embodied Signs of Elite Distinction: Franco-Mauritians' White Skin-Colour in the Face of Change." *Comparative Sociology* 10 (4): 548–570. doi:10.1163/156913311X590628.

Sandbrook, R., M. Edelman, P. Heller, and J. Teichman. 2007. *Social Democracy in the Global Periphery: Origins, Challenges, Prospects*. Cambridge: Cambridge University Press. doi:10.1017/CBO9780511491139.

Scott, J. 2001. *Power*. Cambridge: Polity Press.

Scott, J. 2008. "Modes of Power and the Re-Conceptualization of Elites." *Sociological Review* 56 (suppl 1): 27–43.

Seekings, J. 2011. "British Colonial Policy, Local Politics, and the Origins of the Mauritian Welfare State, 1936–1950." *Journal of African History* 52 (2): 157–177.

Shaw, W. H. 2003. "They Stole Our Land: Debating the Expropriation of White Farms in Zimbabwe." *Journal of Modern African Studies* 41 (1): 75–89. doi:10.1017/S0022278X02004159.

Simmons, A. Smith. 1982. *Modern Mauritius: The Politics of Decolonization*. Bloomington, IN: Indiana University Press.

Storey, W. K. 1997. *Science and Power in Colonial Mauritius.* Rochester, NY: University of Rochester Press.

Truth and Justice Commission. 2011. *Volume 1: Report of the Truth and Justice Commission.* Mauritius: Government Printing.

Vaughan, M. 2005. *Creating the Creole Island: Slavery in Eighteenth-Century Mauritius.* Durham and London: Duke University Press.

Vogt, E. A. 2005. "Ghosts of the Plantation: Historical Representations and Cultural Difference among Martinique's White Elite." PhD diss., University of Chicago.

Index